Orchard and Estate Management

About the Authors

Prof. Bhimasen Naik (b.1961) hails from an agri-horticultural family of Western Odisha (Village: Jamtalia, Tehsil: Sundargarh, District: Sundargarh). He obtained B.Sc. (Ag. & A.H.) and M.Sc. (Ag.) in Genetics and Plant Breeding from Chandra Shekhar Azad University of Agriculture and Technology, Kanpur in 1981 and 1983, respectively, and Ph.D. from Utkal University, Bhubaneswar in1999. He has been awarded ICAR Junior Research Fellowship and CSIR Senior Research Fellowship during his studies. He is a Fellow of the Indian Society of Genetics & Plant Breeding, New Delhi; and life member of the Crop Improvement Society of India, Ludhiana; the Indian Society of Plant Breeders, Coimbatore; the Society for Advancement of Rice Research, Hyderabad; the Indian Society of Oilseeds Research, Hyderabad and the Society for Plant Biochemistry & Biotechnology, New Delhi. He has worked on breeding and genetics of ginger, turmeric, mungbean and linseed, and presently working on rice. He has 36 years of research and teaching experience, and is presently working as Professor of Plant Breeding and Genetics at Regional Research and Technology Transfer Station of the Odisha University of Agriculture and Technology at Chiplima, Sambalpur, Odisha. He has 56 research papers published in peer-reviewed national and international journals, 23 abstracts of symposia, three books, one book chapter (Springer-Nature) and 20 popular articles to his credit. He has guided one M.Sc. (Ag.) and one Ph.D. student. He is associated in the development of five cultivars in ginger (Suruchi), turmeric (Surama, Ranga and Rashmi) and linseed (Arpita); collaborated in development of three cultivars in rice (Ashutosh, Gobinda and Hasanta) and registered one genetic stock of mungbean (BSN 1, INGR 00011).

Dr. Ranjan Kumar Tarai (b.1977) obtained his B.Sc. (Ag.) in 1999 from College of Agriculture, Chiplima under Odisha University of Agriculture and Technology; M.Sc. (Hort.) in Fruits and Orchard Management and Ph.D. (Hort.) in Fruits and Orchard Management in 2002 and 2006 respectively from Bidhan Chandra Krishi Viswa Vidyalaya, Mohanpur, Nadia, West Bengal. He successfully qualified NET in Fruit Science during 2004 conducted by ASRB, New Delhi. He started his career in 2006 as Subject Matter Specialist (Horticulture) at Krishi Vigyan Kendra (Gajapati) and later on joined as Junior Scientist (Horticulture) at Regional Research and Technology Transfer Station, Bhawanipatna and as Programme Coordinator at KVK, Kalahandi under Odisha University of Agriculture and Technology, Bhubaneswar. Dr. Tarai is basically a Pomologist inclined to fruit research and teaching. He has an experience of over 14 years of teaching, research and extension. He has published more than 53 research papers in national and international journals, 05 Books, 26 popular articles, 18 book chapters, 20 extension booklets & Newsletters to his credit. He has guided two M.Sc. Ag. (Hort.) students. He is a life member of several societies in the field of Horticulture. He has devoted his works in development and transfer of technologies in fruits (especially minor fruit crops), vegetables and horticulture-based farming system suitable for resource poor area. Currently he is working as Associate Professor (Fruit Science) at College of Horticulture, Chiplima under Odisha University of Agriculture and Technology, Bhubaneswar.

Dr. Ajit Kumar Sahoo (b.1988) born at Jajpur and did his Graduation from College of Agriculture, Odisha University of Agriculture and Technology, Bhubaneswar in 2011, M. Sc. (Pomology) at Dr. Punjabrao Krishi Vidya Peeth, Akola in 2013 and Ph.D. (Pomology) from Odisha University of Agriculture and Technology, Bhubaneswar in 2017. He worked as Asst. Professor at College of Horticulture, Chiplima under Odisha University of Agriculture and Technology from 2015 to July 2020. Presently he is working as Asst. Professor in AICRP on Palm, Bhubaneswar. He is actively involved in teaching, research and extension activities. He is working on micropropagation of pineapple and canopy management in sapota. He has published 3 books, 2 book chapters, 1 popular article, 17 research papers and 2 review papers in national and international journals/ scientific magazines.

Dr. Purandar Mandal (b.1978) obtained his B.Sc. (Ag.) in 2001 from Veer Bahadur Singh Purvanchal University, Jaunpur, U.P.; M.Sc. (Ag.) Horticulture from Narendra Deva University of Agriculture and Technology, Faizabad, U.P. in 2003; qualified National Eligibility test for Asst. Professorship/lectureship recognized by ICAR/CSIR in Horticulture (Fruit Science) in 2005; Ph.D. (Horticulture) from Odisha University of Agriculture and Technology., Bhubaneswar in 2009. He started his career in 2006 at Reliance Retail Ltd, Bhubaneswar as Senior Associate (Horticulture supply chain), joined at College of Horticulture, Odisha University of Agriculture and Technology, Chiplima, Sambalpur as Assistant Professor (Horticulture) in 2009. Dr. Mandal is basically inclined to post-harvest management, orchard management and teaching. He has an experience of over 12 years of teaching research and orchard management; worked as Principal Investigator in the All India Network Research Project on Onion and Garlic (ICAR) from 2013 to 2018. He has published more than 35 research papers in national and international journals, 03 practical manuals (Fundamentals of Food Technology, Weed Management in Horticultural Crops, Introductory Agro-forestry) for B.Sc. Hort. students, 15 popular articles, 01 book, 05 book chapters to his credit. He has guided two M.Sc. Ag. (Hort.) students and one Ph.D. (Hort.) as a member. He is a life member of several societies in the field of Horticulture. Currently he is working as Assistant Professor (Horticulture) at College of Horticulture, Chiplima.

Dr. Sunil Samal (b.1974) obtained his B.Sc. (Ag.) from Shri Shahu Ji Maharaj University, Kanpur in 1997; M.Sc. (Ag.) in Horticulture from Odisha University of Agriculture and Technology, Bhubaneswar in 1999 and Ph.D. from Utkal University, Bhubaneswar in 2003; subsequently qualified NET in Fruit Science in 2004 conducted by ASRB, New Delhi. He started his career as Training Associate (Horticulture) at Krishi Vigyan Kendra, Dhenkanal in 2005 and later on joined as Programme Coordinator at Krishi Vigyan Kendra, Koraput. In 2010, he was selected for the post of Associate Professor (Post Harvest Management) and joined the College of Horticulture, Chiplima. In teaching he remained for eight years. Besides teaching he worked for the development of Horticulture, Horti-based farming system in the district. To his credit he accumulated an experience of over 15 years of Teaching, Research and Extension. He has published more than 12 research papers in National and International journals, 10 popular articles, 6 Abstracts of Symposia, 11 Extension Booklets and Newsletters to his credit. During his service carrier he attended a number of Refresher courses/Symposia/Seminars conducted by National and State level organizations. Presently, he is working as Senior Scientist (Horticulture) at the Regional Research and Technology Transfer Station, Dhenkanal.

Orchard and Estate Management

Bhimasen Naik
Professor of Plant Breeding and Genetics
Odisha University of Agriculture and Technology
Chiplima Campus, Sambalpur-768 025, Odisha

Ranjan Kumar Tarai
Associate Professor of Fruit Science
Odisha University of Agriculture and Technology
Chiplima Campus, Sambalpur-768 025, Odisha

Ajit Kumar Sahoo
Assistant Professor of Horticulture
Odisha University of Agriculture and Technology
Bhubaneswar Campus, Bhubaneswar -751 003, Odisha

Purandar Mandal
Assistant Professor of Horticulture
Odisha University of Agriculture and Technology
Chiplima Campus, Sambalpur-768 025, Odisha

Sunil Samal
Associate Professor of Horticulture
Odisha University of Agriculture and Technology
Mahisapat Campus, Dhenkanal-759 001, Odisha

NEW INDIA PUBLISHING AGENCY
New Delhi-110 034

NEW INDIA PUBLISHING AGENCY
101, Vikas Surya Plaza, CU Block, LSC Market
Pitam Pura, New Delhi – 110 034, India
Email: info@nipabooks.com
Web: www.nipabooks.com

For customer assistance, please contact
Phone: + 91-11-27 34 17 17 Fax: + 91-11- 27 34 16 16
E-Mail: feedbacks@nipabooks.com

ISBN: 978-93-90591-36-7

Composed and Designed by NIPA.

To
Abhipsa
Sonali
Sanskruti
Late Mrutyunjay Mandal
Dr. Ananta Prasad Samal

Bhimasen
Ranjan
Ajit
Purandar
Sunil

Preface

The Fifth Deans' Committee of Indian Council of Agricultural Research has recently revised the syllabus of B.Sc. (Hons.) Horticulture which is uniform throughout the country. The course 'Horticulture Estate Management' is taught in the sixth semester. The present textbook covers the entire syllabus in 13 chapters. Simple and lucid language has been used for easy understanding of the beginners. The information contained in the book has been gathered from various published sources and internet websites which are mentioned at the end of each chapter under references. Attempts have been made to provide latest information; still some valuable information might have been missed. Questions are set at the end of each chapter to assess the understanding of the students.

We have tried our best to remove the errors, typographical or otherwise, from the text; still there might be some. We would highly appreciate if it is brought to our notice for rectification in next edition. We cherish the encouragement and cooperation received from our family members during preparation of the manuscript. We congratulate M/S New India Publishing Agency, New Delhi for their support and publishing in a short time.

Though the book is primarily written for B.Sc. (Hons.) Horticulture students, the counterparts of B.Sc. (Hons.) Agriculture also may be benefitted. It may serve as a help book for post-graduate students.

We seek suggestions from learned teachers and students for further improvement of the book.

Chiplima **Authors**

Contents

1

Orchard and Estate Management An Overview

1.1 Introduction

Orchard is an area of land on which fruit trees are grown. An orchard is purposeful planting of fruit trees that is maintained for food production. It encompasses fruit or nut-producing trees which are grown for commercial production. Orchard management refers to management practices planned in the fruit orchard to maximize production of high-quality fruit and increase grower profitability. Decisions on soil preparation, crop cultivars, orchard systems and designs, location and site selection, making provision for water availability, canopy management, nutrient management, orchard floor management, pest and disease management, harvesting and mechanization etc. need to be taken care of while establishing an orchard. It is easy to establish an orchard considering its features critically but difficult to manage afterwards. It needs high managerial skills and technically skilled manpower to carry out different orchard operations. Orcharding is a long-term investment. Fruit plants once planted remain committed to the land for many years. Being perennial in nature, any mistake committed during initial years of orchard planning, is very tough to rectify.

Orchards have played an important role in providing fresh nutritious fruits which are rich sources of vitamins and minerals. In modern times, community orchards have revitalization as people rediscover the benefits and pleasures of growing fresh fruit from trees. Once an orchard is established, it can provide a good yield of fruit year after year for decades, requiring little human input when compared to the growing, planting, weeding, watering, pest control involved in annual vegetables. A healthy, diverse and resilient systems can be created while reducing our impact on climate change and environmental degradation. Orchards nearer to cities and towns may encourage people to interact with nature either through becoming custodians of the trees or by simply being around them and enjoying their shade, beauty and fruit. They reduce heating and cooling costs, reduce pollution, take up carbon dioxide, produce oxygen, provide habitat for wildlife, hold water and reduce soil erosion. Sedentary lifestyles and poor diets cause rising rates of diabetes and obesity. Hence, fresh, nutritious fruit and the activity involved in growing and harvesting in the orchard can overcome the above problems. The green environment of an orchards area is a suitable playing area for children to express their physical competences as well as socialization ones. They allow children to explore and discover the elements of nature through physical and social participations (Fjortoft and Sageie, 2000).

1.2 Factors Affecting Orchard Management for Fruit Production

1.2.1 Location and Site Selection

When choosing a potential site for an orchard, issues relating to the land, such as soil type, topography, aspect, slope and irrigation supply need to be considered carefully. The location for orchard should be close to a market and well-established fruit growing region to get easily field labourers, and also to get benefit of experience of other fruit growers and to sell the produce through co-operative fruit grower organizations. All technical inputs, skilled labour, specific equipment should be easily accessible to such sites. The selected site should be free from different biotic and abiotic stresses like pests, diseases, water and nutrient deficiencies and natural calamities like cyclones, flood etc. and sunlight should be amply penetrated to the inner canopy for production of quality fruits. There must be good transport facilities by road or rail for early disposal of perishable fruits.

1.2.2 Climate

Fruit plants have a different type of adaptability to different types of climatic conditions. As per climatic adaptability, fruit plants have been classified in to three: tropical, sub-tropical and temperate types (Table 1.1). Selection of particular type of climate for particular fruit plant is necessary for successful cultivation of fruits (Singh, 2018).

Table 1.1: Fruit crops suitable for different climatic conditions

Tropical climate	Sub-tropical climate	Temperate climate
Mango, Banana, Papaya, Pineapple, Guava, Sapota, Aonla, Custard apple, Carambola, Jack fruit, Grape, Citrus (mandarin, Sweet orange, Acid Lime)	Mango, Citrus (Mandarin, Sweet orange, Lime, Lemon, Grape fruit), Grape, litchi, Avocado, Passion fruit, Aonla, Pomegranate, Date palm, Loquat, Fig	Apple, Almond, Walnut, Pear, Peach, Plum, Apricot, Strawberry, Cherry, Kiwi fruit, Chest nut, Pecan nut, Pistachio nut and Hazel nut

Source: Usha *et al*. (2015).

1.2.3 Soil

The soil best adapted to general fruit culture may be classified into gravelly loam, sandy loam, strong black loam and clay loam. Very light sandy soil, wet swamp and peat soil or dry sand, are generally to be avoided, although there are exceptions. The soil serves as a base for growing plants. The soil of the orchard should be deep and well drained, free from water logging, neutral in reaction (pH 7.0) and should have good water holding capacity (Singh, 2018). Problematic soil like eroded, gullied, saline, alkaline soil should be reclaimed.

1.2.4 Planning

Careful planning is necessary for efficient economic management of orchard. Optimum spacing should be provided to accommodate maximum number of fruit trees per unit area. At least 10 % of the total area should be left for buildings, roads, paths, storehouse, tube well, water channel etc. The farm building or office should be located at the centre. The watch post should be at entry gate. Roads of 10-15 feet width should be provided and each should cross at 90° angle. Provisions should be there for store house, packing house etc. Irrigation system should be properly planned. Deep wells should be dug at convenient places at a rate of one for every 2 to 4 hectares of land. In the absence of bore well a deep farm pond with adequate water storage capacity should be constructed to reserve water from canal or rain for assured irrigation facility. For efficient utilization of every drop of water drip irrigation facility should be provided. Permanent fencing with brick wall, iron pole and barbed wire should be there to protect from stray animals, wild animals and thieves. In absence of permanent fencing temporary fencing of three closer rows with live thorny plants having dense foliage, easy to raise from seeds, quick growing and drought resistant capacity can be preferred. From few years back construction of solar fencing is getting momentum. Planting of 2 alternate rows of wind-break plants like Jamun, Neem, Sisoo, Karanja, Deodar, *Eucalyptus* at 5m interval in north-west side to protect the orchard from hot and cold wave. Sufficient space should be maintained between row of wind break and first row of fruit plant by digging a trench of 90 cm deep at a distance of 3 m from the wind breaks. To avoid competition with fruit plants, roots of wind breaks should be pruned around the trench area and trench filling with pruned materials in every three to four years. Each kind of fruit plants should be assigned to separate blocks. There should be 7.5 cm gradient for each 30 m distance of irrigation channel. Fruits ripening at the same time should be grouped together. In case of dioecious fruit plants, every 3rd tree in every row should be a pollinator. Short growing trees should be allotted to the front and the tall trees at the back for easy watch and ward and beauty. Plant evergreen trees in the front and deciduous plants at back rows. Fruits and berries which are attracting birds and animals should be close to the watchman shed.

1.2.5 Layout

Layout means locating the position of trees, roads and buildings in the orchard being established whereas systems of layout refer to the orderly ways of planting the trees. Any method of layout should aim at providing maximum number of trees per hectare, adequate space for proper development of the trees and ensuring convenience in orchard cultural practices. The various layout systems used are described as below:

a. **Square System:** In this system, trees are planted on each corner of a square whatever may be the planting distance. This is the most commonly followed system, easy to layout and permits inter-cropping and cultivation in two directions. The central place between four trees may be advantageously used to raise short lived filler trees. It is simplest and most common for planting of orchards. Only defect is presence of vacant space in the centre till the plants grow up sufficiently.

b. **Rectangular System:** In this system, trees are planted on each corner of a rectangle. The distance between any two rows is more than the distance between any two trees in a row. The wider row spaces permit easy intercultural and mechanical operations. It is adopted for planting of those fruit trees which require less distance between plant to plant than row to row distance. It has almost all the advantages of square system but cultivation between plant to plant is difficult.

c. **Hexagonal System:** In this method, the trees are planted on each corner of an equilateral triangle. This way six trees form a hexagon with the seventh tree in the centre. Therefore, this system is also called as 'septule' as a seventh tree is accommodated in the centre of hexagon. This system provides equal spacing but it is difficult to layout for doing cultural operations. This system accommodates 15% more trees than the square system.

d. **Diagonal or Quincunx System:** This is the square method but with one more plant in the centre of the square. This will accommodate double the number of plants, but does not provide equal spacing. The central filler tree chosen should be a short lived one and should be removed after a few years when main trees come to bearing. This system can be followed when the distance between the permanent trees is more than 10m. It is advantageously adopted, only where the maximum use of the land is desired, especially on fertile soils.

e. **Triangular System:** Triangular system is based on the principle of isolateral triangle. The distance between any two adjacent trees in a row is equal to the perpendicular distance between any two adjacent rows. When compared to square system, each tree occupies more area and hence it accommodates few trees per hectare than the square system.

f. **Contour System:** It is generally followed on the hills with undulated topography to minimize land erosion and to conserve soil moisture. In this system, plants are planted along the contour across the slope. The planting distance under the contour system may not be uniform. Trees will be in a straight line from one side only. Rows are represented by lines passing through the same contour.

DIFFERENT SYSTEMS OF LAYOUT

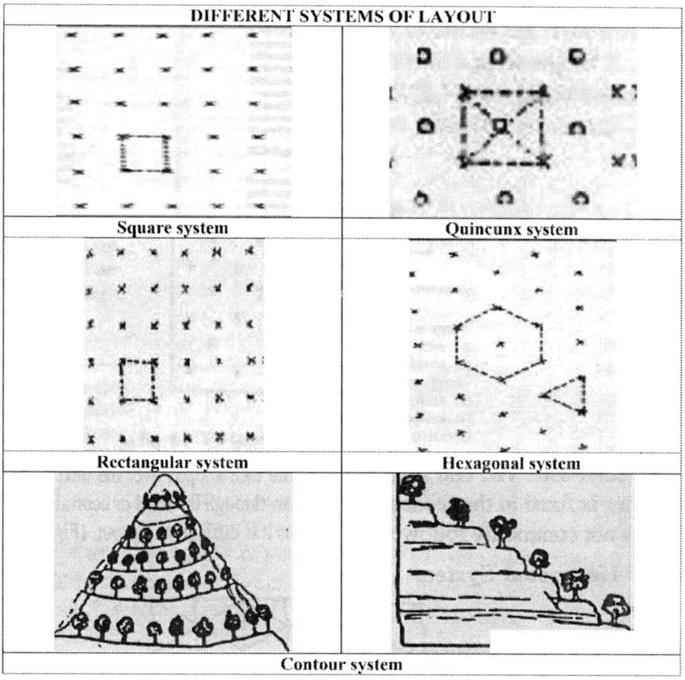

1.2.6 Planting

Planting time depends upon the locality, type of fruit crops to be grown, season of growing, soil type etc. Under Indian condition there are two seasons of planting, i.e., monsoon planting during June to August and spring planting between February to March. Crops suitable to monsoon planting are citrus, mango, sapota, guava and all others can be planted during spring. Pits are dug in the field marked earlier at least one month before as per the system of planting using planting board. Planting board is a wooden plank of 1.5 m long,

10 cm wide and 2.5 cm thick. Pits of size 45 to 90 cm^3 are opened as per the seedling to be transplanted. Pits are left open for about 15-30 days for proper solarisation. During planting application of 1 kg neem cake, 50 g bone meal per pit along with 1 basket of FYM may be applied. During planting care should be taken not to bury the graft union; it should be slightly above the ground. After the planting hole is filled, irrigate immediately to saturate the soil. After soil is settled, the graft union should be ensured to be placed at least 2 - 3 inches above the soil level (Table1.2).

Table 1.2: Suggested cultivars and planting distance in different fruit crops

Fruit crop	Cultivar	Planting distance (m × m)
Aonla	NA-6, NA-7, Kanchan, Krishna, Goma Aishwarya	6 × 6
Apple	Red chief, McIntosh, Golden Delicious, Red Delicious	4 × 4 to 5 × 5
Apricot	New Castle, Shipley Early, Turkey, Charmagaz	6 × 6
Banana	Robusta, Dwarf Cavendish, Grand Naine, Rasthali	2 × 2
Ber	Gola, Seb, Goma Kirthi, Umran, Banarasi Karaka	7 × 7
Custard Apple	Balanagar, Mammoth, Atemoya, Arka Sahan	5 × 5
Date Palm	Halawy, Khalas, Khuneizi, Barhee	10 × 10
Fig	Capri fig, Smyrna fig, White San Pedro, Conadria, Excel	4 × 4
Grape	Anab-e-Shahi, Thompson Seedless, Arka Vati for South India and Perlette, Pusa Urvashi for North India	3 × 3
Karonda	Green, pink, white	2 × 2
Kiwi fruit	Pistillate: Abbott, Bruno, Hayward, Monty Staminate: Tomuri, Allison, Matua	6 × 6
Lemon	Kagzikalan, Nepali oblong, Lisbon, Eureka, Villafranca	5 × 5
Lime	Parmalini, Vikramand, Rasraj	5 × 5
Litchi	Swarna Roopa, Shahi, China	10 × 10
Loquat	Tanaka, California Advance, Pale Yellow, Golden Yellow	5 × 5
Mandarin	Kinnow, Nagpur Mandrin	4.5 × 4.5, 6 × 6
Mango	Amrapalli, Arka Aruna, Dashehari, Pusa Prathibha, Pusa Arunima, Pusa Surya, Pusa Shreshth, Pusa Lalima, Pusa Peetamber	3 × 3 to 5 × 5. Other vigorous cultivars (10 × 10)
Papaya	Pusa Nanha, Pusa Dwarf, Surya, Arka Prabhat	1.25 × 1.25
Passion fruit	Purple,Yellow, Kavery	2 × 2
Peach	Red heaven, Floridasun, Shan-e-Punjab (Low chilling)	4 × 4
Pear	Baghugosha, Punjab Nectar (Low chilling)	5 × 5
Phalsa	Dwarf Type	2 × 2 m

Fruit crop	Cultivar	Planting distance (m × m)
Pineapple	Queen, Mauritius, Kew	30 cm × 60 cm × 90 cm
Plum	Mariposa, Santa Rosa, Kelsey, Kala Amritsari	4 × 4
Pomegranate	Bhagwa, Mridula, Ganesh, Jyothi	3 × 3
Sapota	Cricket Ball, PKM-1, PKM-3	8 × 8
Strawberry	Chandlar, Pusa Early Dwarf, Tribute	40 cm × 25 cm
Sweet orange	Blood Red, Hamlin, Pineapple, Jaffa, Valencia Late	6 × 6

Source: Usha *et al.*, (2015)

1.2.7 Nutrient Management

Nutrient management, particularly in fruit crops, is the skillful task of matching a specific field, soil, climate, and crop management conditions to rate, source, timing, and place of nutrient application (Mikkelsen, 2011). Integrated nutrient management refers to the use of chemical fertilizers in conjunction with organic manures, green manures, crop residues and legumes in a cropping system and locally available resources with objectives of sustaining high yield and ensuring environmental safety. The selection of appropriate nutrient management practices for individual horticulture farms needs to be tailored to the specific conditions existing at a site. The science and art of nutrient management has progressed immensely during the last decade because of the integration of new developments in the areas of computers, simulation models, GIS, GPS, and other *in situ* sensors. Technology will continue to advance and help integrate these multiple factors for agricultural sustainability and profitability (Ganeshmurthy *et al.*, 2016) (Table 1.3 and Table1.4).

Table 1.3: Manuring and fertilizer application in important fruit crops

Crop	Time of application
Mango	The nutritional requirement of mango varies with the region, soil type and age. A dose of 73g N, 18g P_2O_5 and 68g K_2O/ year of age from first to tenth year and thereafter a dose of 730g N, 180g P_2O_5 and 680g K_2O should be applied in 2 split doses during June–July and October respectively. Spraying of zinc sulphate (0.3%) during February, March and May is recommended to correct the zinc deficiency. Spraying of Borax (0.5%) after fruit set twice at monthly intervals and 0.5% manganese sulphate after blooming corrects boron and manganese deficiencies respectively. Organic manures and phosphatic fertilizers should be applied immediately after harvest, whereas ammonium sulphate should be given before flowering.

Crop	Time of application
Banana	Banana is a heavy feeder crop that requires very large quantity of nutrients for growth and yield, which accounts for 20–30% of the total cost of production. For normal plant growth and development 100–250g of N/plant is advised depending on nutrient status of soil and cultivar. Urea is commonly used as a source of N. It should be applied in 3–4 splits. Application of 150g N in vegetative phase and 50g N in reproductive phase enhances the yield and delays the leaf senescence. Application of 25% N as farmyard manure and 1kg neem cake is beneficial. Phosphorus is applied in single dose at the time of planting and quantity of P_2O_5 depends upon soil type and varies from 20 to 40g/plant. Potassium is indispensable in banana nutrition due to its role in vital functions. It is not stored and its availability is influenced by temperature. Thus, continuous supply is required to be assured at finger-filling stage. Application of K (100g) in 2 splits during vegetative phase and 100g in 2 splits during reproductive phase is recommended. Application of 200–300g K_2O is recommended depending upon cultivar. Invariably, plantains require higher K than other group of cultivars. Muriate of potash is invariably used as source of K. But in soils with pH above 7.5, potassium sulphate is advantageous. Among micronutrients, Zn, Fe, B, Cu and Mn play an important role in normal growth and development of banana. The application of Zn (0.1%), B (0.005%) and Mn (0.1%) improves yield.
Mandarin Orange	Application of 20–25kg farmyard manure together with 0.4kg calcium ammonium nitrate is recommended at the time of planting. A mixture of 90g each of N, P, K/tree may be applied in first year after planting. This dose may be gradually increased to 450g each of N and P, and 900g of K/tree in the seventh year and kept constant thereafter. The micronutrients should be supplied through foliar spraying (multiplex micronutrient mixture @3 ml/l) during November and March. Normally young plants are manured once a year, while bearing plants more than once. Total amount of P and K fertilizers is applied at one time, while N fertilizers are applied in 2 or 3 split doses. In north-western India, manures are added twice a year, once during June and another after harvesting in December–January. In central India, mandarins are manured during December for ambe bahar and in May for mrig bahar. However, for Coorg mandarin, manuring thrice (March–April, June–July and September–October) is highly beneficial. For soil application, first 30–40cm area around the tree trunk should be given hoeing and fertilizers should be applied in the form of ring. They should be mixed thoroughly. The trees should be irrigated immediately
Guava	For guava-growing regions of the country, different fertilizer schedules—600g N, 400g P and 400g K in northern region; 260g N, 320g P and 260g K in eastern region; 900g N, 600g P and 600g K in southern region and 600g N, 300g P and 300g K/plant/year in western region—have been recommended. In north India, fertilizer is given in the first week of May for rainy season crop and in first week of July for winter season crop. In West Bengal, fertilizers are applied in 2 equal split doses, one in January and the other in August. At Bangalore, full K and 70% N are applied in June and full P and 30% N in September. Since 48% of feeder roots of guava are found in the surface soil up to 25cm depth, the fertilizer should be placed in 25cm trenches 1m away from the trunk for better uptake. Bronzing can be reduced by improving the soil pH and treating the soil with N, P, K and Zn at 200, 80, 150 and 80g/year respectively, or fortnightly foliar spraying of these nutrients each at 2% for 4 months.

Crop	Time of application
Papaya	Papaya is a heavy feeder and needs heavy doses of manures and fertilizers. Apart from the basal dose of manures applied in the pits, 200–250g each of N, P_2O_5 and K_2O are recommended for getting high yield. Application of 200g N is optimum for fruit yield but papain yield increases with increase in N up to 300g. A dose of 250g N, 250g P and 500g K/plant is recommended for papaya Coorg Honey Dew under Bangalore conditions, while 200g each of N, P and K in split doses in the first, third, fifth and seventh month is recommended for papaya Co 1 under Coimbatore conditions. Deficiency of lime and B has often been observed in papaya orchards. Spraying of 0.5% zinc sulphate (twice) and one spray of Borax (0.1%) may be done depending upon the nutrient status of soil.

Source: Chadha, K.L (2001)

Table 1.4: Nutritional requirement of full grown up (other fruit) trees/plant/year

Crop	Manures (Kg)	N:P:K (g)	Time of application
Ber	50	750: 150: 150	Half in July and half in October
Date Palm	50	1200: 350: 300	March-April
Pomegranate	50	625 :250: 250	Dec-Jan: Ambe Bahar May-June: Mrig Bahar
Aonla	50	800:300: 400	Half N, half P and full K in Jan-Feb and rest half N and half P in Sep
Phalsa	10	100: 50: 25	After pruning
Fig	25	900: 150: 200	February-March
Guava	40	360: 180: 180	Half N, Full P and full K in July-August, and half N in October

Source: More and Sharma (2008)

1.2.8 Water Management

An efficient water management refers to artificial application of water *i.e.* irrigation in crop root zones in case of soil moisture deficit and removal of water *i.e.* drainage from the root zone in case of excess so as to provide the crops a most optimum soil moisture regime for best production. It means how best we are in irrigating and dewatering our fields so that plants are protected from stress as well as water logging (Goswami, 2015). When rainfall is not adequate, irrigate the newly planted trees at least once each week during the first growing season. Earthing up of soil around the tree trunks can help prevent spread of some fungal diseases. Some trees like mango, citrus require stress period for flowering; hence irrigation is avoided prior to flowering. Irrigate at 10-15 days intervals when fruits are growing till maturity.

1.2.9 Training and Pruning

Training is an operation by which the plant is made to develop an orderly structure or frame work which is done by staking, tying, supporting, propping, trailing or spreading on certain structure with or without pruning of plant. Basic training types are central leader system, open centre system and modified leader system. Specific types of training are bush system, pyramid system, espalier system, cordon system, tatura trellis, head system, kniffin system, telephone system and bower system. Pruning refers to judicious removal of plant parts like roots, leaves, stems, branches, flower buds, flowers, even immature fruits for good yield and quality fruits. Different types of pruning are heading back, thinning, ringing or girdling, notching, nicking, root pruning, smudging, bending, coppicing, pollarding, lopping, pinching, disbudding, containment pruning and alternate differential pruning.

1.2.10 Weed Management

Weeds can compete with fruit trees for both water and nutrients. Weed competition in young fruit trees reduces tree growth and efficiency, and therefore decreases fruit production and fruit size (Merwin and Stiles, 1994). Fruits have a great potential but orchards are generally, infested with various types of annual, biennial and perennial weeds which compete with the fruit plants for nutrients, space, moisture, light and adversely affect growth, vigour, fruit set, flower initiation, yield, fruit quality and winter hardiness (Majek *et al*., 1993) and thereby directly reducing the productivity of fruit trees. Since most horticultural crops are very slow in growth, especially at the early stages of their establishment, herbicides become imperative to begin weed control early enough in order to ensure high yields and quality (Njoroge,1999). Weed management is a system approach whereby whole land use planning is done in advance to minimize the invasion of weeds in aggressive forms and give crop plants a very strong competitive advantage over the weeds. The system approach is called integrated weed management (IWM).

1.2.11 Cropping System

In the cropping systems, a number of crops are grown together or they are grown separately at short intervals in the same field (Rana and Rana, 2011). It may also be defined as "a pattern of growing a compatible crop or its combination followed on a particular farm year after year in order to get maximum utilization of land in time and space, in addition to the practices and technologies with which the crops are produced". In case of multispecies cropping, there is large number of compatible crops which are grown in a same unit of land. There are a number of multispecies cropping system followed and some of them are: a) Inter-cropping, b) Mixed cropping c) Sequential cropping d) Relay cropping and e) Multi-storey cropping (Table 1.5).

Table 1.5: Some fruit crop based cropping system with their intercrops

No.	Crop	Main crop with intercrops	References
1.	Aonla	Aonla + Fenugreek + chick pea + mustard + cumin (Rabi season); Moth bean (Kharif season)	Awasthi *et al.* (2009)
		Aonla + Ber + Karonda + Cluster bean + Brinjal; Aonla + Ber + Karonda + Moth bean + Mustard	Arya *et al.* (2010).
2.	Ber	Ber (Jharber and Bordi) + Mung, Cluster bean, Pearl millet, Moth bean, Gram, Vegetable crops	Faroda (1998)
		Ber (*Zizyphus mauritiana*) + Green gram	Gupta *et al.* (2000)
		Ber (cv. Gola) + Cluster bean+ Mustard + Indian aloe; Wheat + Groundnut	Saroj *et al.* (2003)
		Ber + Wheat; Ber + Chick pea; Ber + Kulthi	Pareek and Nath (1996); Kumar *et al.* (2005)
3.	Mango	Mango + *Phaseolus acutifolius* Genotype "Frijol Escumite" + *Cajanus cajan* (Pigeon Pea) + *Phaseolus vulgaris* Genotype "Frijol de Vara"	Agreda *et al.* (2006)
		Mango + Guava + Cowpea Guava + Groundnut + Black gram	Nath *et al.* (2007)
4.	Mandarin	Mandarin + Ginger	Upadhyay and Subba (1994)
		Mandarin + Pea	Ghosh *et al.* (1985)
5.	Lemon	Lemon + French bean	Vanlalhruaia *et al.* (2013)
6	Banana	Banana +Sweet potatoes	Bekunda and Woomer (1996)

1.2.12 Pollination

For successful cropping effective pollination plays a significant role. The plant may be self-pollinated or cross-pollinated. Various factors that affect pollination are dioecious nature of plant, peculiar structure of flowers, dichogamy, functionally unisexual flower, self-incompatibility, requirement of artificial pollination, impotence of pollen grains, abortive flowers etc. Cropping also depend upon fruit set. The plant may be fruitful one or unfruitful or barren. Unfruitfulness is associated with some external factors like nutrient supply, pollinisers, training and pruning, locality, season, temperature, light, disease and pest, unwanted spraying of chemicals and fungicides during blooming. For effective pollination, many fruit and nut trees require a different cultivar (variety) or, in some cases, a different species from the same plant family. Most fruit trees are pollinated by bees, house flies or other insects and to ensure good yields bee hives may be needed to be placed in the orchard. Most nut trees are pollinated by wind and they do not need bees for pollination. The number of pollinators may vary from 5% to 50% depending on the species and cultivar. Some cultivars are self-fertile and do not need separate pollinator trees.

1.2.13 Orchard Soil Management

Orchard soil management aims at conserving soil moisture, soil nutrition, controlling soil erosion, little disturbance to soil and reduction in cost of cultivation. Most of the soil management practices in the orchard enhance nutrient and water supply needed for growth and production of fruit trees.

1.2.13.1 Clean Culture

It refers to regular and deep ploughing to remove weeds in the orchard. Clean cultivation though keeps the land weed-free for more days in a year, it brings lot of problems like depletion in humus content, injury to feeding roots, depletion in nitrogen content due to leaching, creation of hard pan and soil erosion. So, avoidance of deep and frequent cultivation should be practised.

1.2.13.2 Sod Culture and Cover Crops

In order to maintain soil fertility status, mowing down of sod grass should be done regularly in addition to adequate nitrogen fertilization. Sod culture usually increases leaf phosphorous concentration in apples. Shallow rooted grasses such as Kentucky bluegrass or annual blue grass deplete less moisture from the orchard soil profile than deep rooted sods such as clover and ryegrass. In summer, regular mowing of grasses is suggested to conserve moisture in soil by reducing water use by grasses. On the other hand, sod improves other physical properties as reflected by decrease in bulk density and increase in soil porosity.

1.2.13.3 Mulching

Organic mulches add significant amount of nitrogen to the soil and consequently improve soil fertility. Large amount of potassium is added to the soil by mulching. It is well recognized that conservation of soil moisture is the most significant advantage of mulching. This useful effect is more pronounced during the drying periods of May-June.

1.2.14 Crop Regulation

Crop regulation is the basis for the regular and quality crop which can be achieved through manual thinning, chemical thinning, selective harvesting, training, summer and winter pruning, prevention of pre-harvest fruit drop, etc.

Usually 30-40 leaves per fruit are sufficient to produce fruit of good quality. Irrigation management during the critical stages of fruit development is essential. Application of growth regulators like GA_3 at 200 ppm and ethrel at 200 ppm delay flowering in spring by 3-7 days depending upon species and thus markedly increase fruit set. Application of GA_3 and ethrel help in better synchronization of flowering period and cross-pollination. In blossom thinning, chemicals are applied at bloom, whereas in fruit-lets thinning, chemicals are sprayed a few days after petal fall, i.e., after fruit set. In mango approximately only 0.1% of perfect flowers develop fruits to mature because of high rate of fruit drop. Extent of fruit drop in mango can be controlled by regular irrigation during fruit development period only after fruit set. Biennial bearing in mango can be controlled by training the plant to open centre system or through application of Paclobutrazol and KNO_3. Smudging enhances fruit set in mango. *Bahar* treatment in guava and pomegranate regulates flowering and fruiting.

1.2.15 Rejuvenation of Old Orchards

The decline in productivity of old and dense orchards existing in abundance has become a matter of serious concern for the orchardists, traders as well as scientists. In India 30-35% area under fruit crops is occupied by old, senile, and pest- and disease-affected orchards. For overcoming the problem of unproductive and uneconomic orchards existing in abundance, large scale uprooting and replacement with new plantations will be a long term and expensive strategy. The rejuvenation technique involves the heading back of older trees showing marked decline in annual production and quality of produce to the extent of 1.0 to 1.5 metre height above the ground level during May-June or December-February with the objective of facilitating production of new shoots from below the cut point and allowing the development of fresh canopy of healthy shoots. The newly emerging shoots are allowed to grow up to 40-50 cm length and then further pruned for emergence of multiple shoots below the pruning point to modify the tree structure and maintain canopy size. Profusely emerging shoots in the inner canopy are also pruned out to promote branching. The multiple shoots developed as a result of second pruning are capable of producing flower buds (Baba *et al.*, 2011). Top working is a practice of converting an undesirable plant to desirable one by budding or grafting the top portion when seedling fruit trees are at active growth stage. In the frame working, almost all the primary scaffold branches are retained on the tree and small laterals arising from secondary scaffold are grafted by side grafting method. It is done at higher locations.

1.2.16 Marketing

The fruits can be sold at the farm gate, at farmer's markets, to processors, direct to supermarkets, through wholesale markets or export markets. Growers may choose to sell their produce themselves or, through cooperatives, agents or growers' organisations.

1.2.17 Post Harvest Management

Post-harvest management and storage of fruit is important to maintain shelf-life and preserve its quality, and is also essential to have access to cool rooms to pre-cool fruit and hold fruit prior to shipment to market. Controlled Atmosphere Storage (CA) can be used to extend the shelf-life of many fruits even further. A combination of modification of temperature, oxygen (O_2), carbon dioxide (CO_2) and relative humidity (RH) levels can extend the shelf-life of fruit for up to 12 months.

1.2.18 Climate Change

Climate change may increase extreme events such as sun burn, lack of chilling and prevalence of frost. The impact of climate change on rainfall and the security of irrigation water supply is one of the key factors to consider for sustainable production of tree crops into the future.

1.3 Types of Orchard

Depending upon the purpose for which they are maintained, orchards are of different types, namely:

1.3.1 Low Density Orchard

Plants on this type of orchard are raised on standard rootstocks, which are vigorous and therefore require more spacing. The number of trees per unit is less and thus named as low-density orchard. It is a traditional system of planting and trees are planted at wider spacing, accommodating about 100-250 plants/ha. Here dwarfing rootstock is not used. Trees acquire commercial production potential after 10-15 years of planting. Output from orchard during early 10-15 years is less. Here less input is used and it is popular among growers.

1.3.2 High Density Orchard

High density orchard was pioneered for temperate fruits in Europe during 1960. High density planting (HDP) technique is a modern method of fruit cultivation involving planting of fruit trees densely, allowing small or dwarf trees with modified canopy for better light interception and distribution and ease of mechanized field operation. In India, HDP has been proved useful in many fruit crops, e.g., pineapple, banana, mango, apple and citrus. The underlying principle of an HDP is to make the best use of vertical and horizontal space per unit time and to harness maximum possible return per unit of inputs (Choudhary *et al.*, 2015). These plants are dwarfed by adopting different horticultural techniques like use of dwarfing rootstocks, genetically dwarf scion cultivars, by adopting special systems of training and pruning, use of plant growth retardants, and other such practices. These dwarf trees are more productive, easy to handle and the income of the growers can be maintained steady. Kerutagi *et al.* (2017) analysed the comparative economics of traditional and high-density mango cultivation in Karnataka and observed that in high density orchard, the average yield obtained was more (7.86 t/acre) than in traditional orchard (3.50 t/acre). Kumar *et al.* (2012) analysed studies on high density planting in almond in Kashmir valley and found that maximum nut yield was recorded in higher plant density (4.01 t/ha) followed by medium density (3.06 t/ha) and minimum in lower plant density (2.54 t/ha). (Table 1.6)

Merits: It induces precocity, higher yield and improves fruit quality. It reduces labour cost resulting in low cost of production. It enables the mechanization of fruit crop production and also it facilitates more efficient use of fertilizers, water, solar radiation, fungicides, weedicides and pesticides.

Demerits: Initially become little costly than conventional system and there may be high incidence of some diseases, e.g., sigatoka leaf spot and fingertip in banana. In high density orchard, the economic life span of the orchard becomes lower and there is chance of reduction in fruit size and weight. Intercultural operation becomes difficult.

Table 1.6: Comparison between traditional low density and high-density orchard

Sl. No.	Attributes	Low Density Orchard	High Density Orchard
1	Tree numbers /ha	Few large trees/ha (150-200)	Many small trees/ha (500-1,00,000)
2	Bearing	After two years	From first year
3	Yield	Lower	Higher
4	Management	Difficult to manage due to large tree size	Easy to manage due to small tree size
5	Labour Requirement	More	Less
6	Cost of Production	Higher	Lower
7	Harvesting	Difficult	Easy
8	Quality	Large canopy, poor sunlight penetration and poor-quality fruits	Small canopy, better air and sunlight penetration and high-quality fruits with good colour development

Source: Choudhary *et al.*, (2015).

1.3.3 Progeny Orchard

The progeny trees are the basic component for the multiplication of fruit plants and ultimately decide the quality and productivity of the fruit crops. Hence, it is necessary to preserve the true-to-type and best quality mother trees in the progeny orchard. The mother plant must have been tested for its performance over a number of years. It must be free from transmittable diseases and be in a healthy condition. The fruit shape, size and quality must conform to the typical specification of the cultivar. The progeny trees should be planted by proper lay out, keeping in view the geographical conditions of the area. The high yielding clones of recommended cultivars should be selected for developing new mother block of the progeny trees. The selected plants should be planted in a closer spacing in separate block in order to get continuous supply of scions. These plants should be severely pruned to keep them in vegetative phase and to produce enough shoots for propagation purposes. The marked trees of outstanding merit for taking bud wood should be kept for bud wood only and no fruit crop be taken from these trees. A map of the progeny orchard and fruit nursery should be maintained indicating the layout of the promising cultivars planted.

The progeny orchards are maintained by nurserymen for maintaining the superior types of fruit trees with all the desirable characters mainly for the purpose of propagation. The progeny orchard may supply bud wood to the fruit nurseries and the orchardists. The plants maintained in these orchards are known as mother plants. The demerits of this type of orchard are: complexity involved in the management of fruit crops which differ in their cultural requirements, spacing, fertilizer and irrigation and the whole of the orchard may not come to harvest at one time and it may be difficult to meet the cultural need of different cultivars.

1.3.4 Experimental Orchard

These orchards are generally managed in the colleges and research institutions with the main plan of conducting different experiments for the benefit of the fruit growers in a scientific way.

1.3.5 Clonal Seed Orchard

Here, seed orchard raised from selected clones are propagated by grafting, cutting, air-layering or tissue culture. This type of orchard is established by the use of grafts, cuttings, air-layered plants, tissue culture plantlets or other methods of vegetative propagation.

1.3.6 Seedling Seed Orchard

Here, seed orchard raised from seedlings are produced from selected parents through natural or controlled pollination. This type of orchard is established from seedlings, i.e. by seed propagation from selected parents.

References

Agreda FM, Pohlan J, Marc J and Janssens J (2006). Effects of Legumes Intercropped in Mango Orchards in the Soconusco, Chiapas, Mexico. *In: Conference on International Agric. Research for Development*, pp. 1-6.

Arya R, Awasthi OP, Singh J and Arya CK (2010). Comparison of fruit based multi-species cropping system under arid region of Rajasthan. *Indian J. Agric. Sci.*, **80** (5): 423-426.

Awasthi OP, Singh IS and More T (2009). *Indian J. Agri. Sci.,* **79** (8): 587-591.

Baba JA, Akbar PF and Kumar V (2011). Rejuvenation of old and senile orchards: A Review. *Annals of Horticulture*, **4** (1): 37-44.

Bekunda M and Woomer PL (1996). *Agric. Ecosys. Environ*, **59** (3): 171-180.

Bhargava BS and Chadha KL (1993). Leaf nutrient guide. In: *Advances in Fruit Crops*, Malhotra Publishers, New Delhi, India.

Chadha KL (2001). *Hand Book of Horticulture*, Directorate of Information and Publication of Agriculture, ICAR.

Chadha K L (2008). *Hand Book of Horticulture.* ICAR, New Delhi.

Choudhary HD, Khandelwal SK, Choudhary MK, Gadawal OP and Choudhary MR (2015). High density and meadow orchard planting system in fruit crops. *Popular Kheti*, **3** (1): 22-27.

Faroda AS (1998). Arid zone research: an overview. In: Faroda AS and Manjeet Singh Ed. *Fifty Years of arid zone research in India.* CAZRI, Jodhpur. pp. 1-16.

Fjortoft I and Sageie J (2000). The natural environment as a playground for children: Landscape description and analyses of a natural playscape. *Landscape and Urban Planning*, **48:** 83-97.

Ganeshmurthy AN, Kalaivanan DD and Selvakumar G (2016). Nutrient management in horticultural crops. *Indian J. Fert.*, **11** (12): 30-42.

Ghosh D, Mitra SK and Bose TK (1985). Fruit growing in India. In: *Proceeding of the 3rd National Citrus Seminar*, Calcutta, India, pp. 1-2.

Goswami AK (2015). Methods of irrigation in fruit crops. In: *Fundamental of Fruit Production,* edited by Usha K, Thakre M, Goswami AK and Nayan Deepak, Division of Fruits and Horticultural Technology, Indian Agricultural Research Institute, New Delhi, pp.70-78.

Gupta JP, Joshi DC and Singh BB (2000). *Management of arid agro-system.* In: *National resource management for agriculture production in India* (Eds. J.S.P. Yadav and G.B. Singh). International conference on managing natural resource for sustainable agriculture production in the 21st Century.

Kerutagi MG, Deshetti MB and Abhilash K (2017). Comparative economics of traditional viz high density mango cultivation in Karnataka. *Asian J. Agric. Extension, Economics and sociology,* **18** (3): 1-12.

Kumar D, Seema J, Batra PK and Gill AS (2005). Indian *Journal of Agricultural Sciences,* **75** (4): 222-224.

Kumar D, Ahmad N and Verma MK (2012). Analyzed studies on high density planting in Almond in Kashmir valley. *Indian J. Hort.,* **69** (3): 328-332.

Kumar N (2010). *Introduction to Horticulture*. Oxford and IBH Publishing Co. Pvt. Ltd., New Delhi.

Majek BA, Neary PE and Polk DF (1993). Smooth pigweed interference in newly planted peach trees. *J. Production Agric.*, **6**: 244-246.

Merwin IA and Stiles WC (1994). Orchard groundcover management impacts on apple tree growth and productivity, and soil nutrient availability and uptake. *J. Amer. Soc. Hort. Sci.*, **119:** 216–222.

Mikkelsen RL (2011). The "4R" Nutrient Stewardship Framework for Horticulture *Hort. Technology*, **21**(6): 658-662.

More TA and Sharma BD (2008). Paper on "*Recent Technologies for Developing Arid and Semi-Arid Horticulture in India.*" Paper presented in the 4th Indian Horticulture Congress held at OUAT, Bhubaneswar, 2008.

Nath B, Yadav MS, Kumar S and Sikka AK (2007). *Acta Hort.*,**735**: 277-295.

Njoroge JM (1999). *17th East African Biennial Weed Science Conference proceedings*, pp: 65-71.

Rana SS and Rana MC (2011). *Cropping System.* Department of Agronomy, College of Agriculture, CSK Himachal Pradesh Krishi Viswavidyalaya, Palampur, 80 pages.

Saroj PL, Dhandar DG, Sharma BD, Bhargava R and Purohit CK (2003). *Indian J. Agroforestry,* **5** (1&2): 30-35.

Singh J (2018). *Fundamentals of Horticulture*. Kalyani Publication, Ludhiana.

Upadhyaya RC and Subba JR (1994). Annual Report 1994. Department of Horticulture, Government of Sikkim, Krishi Bhawan, Tadong, Sikkim.

Usha K, Thakre M, Goswami AK and Nayan Deepak G (2015). In: *Fundamental of Fruit Production,* Division of Fruits and Horticultural Technology, Indian Agricultural Research Institute, New Delhi, pp.6-33.

Vanlalhruaia H, Gopichand B and Lalnunmawia F (2013). *Sci Vis.*, **13** (1): 40-44.

e-resources

http://ecoursesonline.iasri.res.in

Outcomes Assessment

Part A

Answer the following questions true or false.

1. Orchard is an area of land on which vegetables are grown. True/False
2. Organic mulches add significant amount of nitrogen to the soil. True/False
3. High density planting has been proved useful in pineapple. True/False
4. The rejuvenation technique does not involve the heading back of older trees. True/False
5. Sod culture usually increases leaf phosphorous concentration in apples. True/False

Part B

Answer the following questions.

1. Differentiate between layout and systems of layout.
2. What is clean culture?
3. What is high density orchard?
4. Most fruit trees are pollinated by — and —.
5. Orchard soil management aims at conserving/depleting soil moisture.

Part C

Write a brief note on each of the following.

1. Discuss the various layout systems of orchards.
2. Compare the low density and high-density orchards.
3. Explain progeny orchard.
4. Nutrient management in orchard.
5. Rejuvenation of old orchards.

2

Clean Cultivation, Sod Culture and Sod Mulch

2.1 Clean Cultivation

This type of cultivation is extensively followed in India. It involves regular ploughing of the land followed by removal of weeds. Tilling the orchard soil regularly provides a weed free environment in the orchard. Traditional tillage with the spade or hoe involves shallow cultivation and partial burying of weeds. Use of animals for ploughing reduces the labour cost but some superficial tree roots get damaged in ploughing. Tractor drawn tool can be used during initial years when the orchard trees are not fully grown. Branches hanging low, spreading in the inter-row spaces hinders the movements of the tractor even in widely spaced rows at maturity in most of the orchards. Under such situations small power tillers can be used. Tillage controls weeds partly through burying weeds and partly through exposing underground parts of weeds to desiccation. The land around the trunk must be cleaned separately with the help of spade as it is difficult to operate a power implement in the vicinity of the trunk. This is very instrumental in saving soil moisture. It has immense effects on the soil properties. Cleaning removes the protecting layer of vegetation and exposes the soil surface to weathering, and thus reduces the frequency of additions of organic matter to soil. Cultivation physically disturbs soil, mixing upper and lower layers. In this management system, the interspace between the trees is kept clean by tillage by removal of weeds. Clean cultivation by tillage influences the soil aggregate size and stability, porosity and pore size distribution, and the development of sub-surface layers of high bulk density (hard pans). The first change that occurs with tillage is the disruption of established soil pore continuity (Chan and Mead 1989). Organic carbon and nitrogen decline to a new equilibrium value depending on tillage intensity, cropping system and climate. Hermawan and Cameron (1994) reported that there was little difference in infiltration rate between tilled and minimum tilled soils. Soil moisture content at the time of tillage is the single most important factor controlling the effect of tillage on the soil structural form. The friability of soil is greatest when tillage is carried out at or just below the soil lower plastic limit (Ojeniyi and Dexter, 1979). Clean cultivation is generally followed in orchards.

2.1.1 Merits

Clean cultivation or tillage may have beneficial effects on productivity of the orchard. It can increase microbial activity by increasing gas exchange (aeration) and water penetration. Increased biological activity can promote the degradation of unwanted coarse organic material in the soil but it also can promote degradation of desirable organic matter causing decrease in the exchange capacity of the soil. Clean cultivation reduces weeds that compete with the crop for light, water and nutrients. It facilitates subsequent orchard operations like spraying, irrigation, harvesting, incorporation of cover crops and fertilizers. Clean cultivation improves aeration and physical condition of the soil. There is improvement in the infiltration of water through breaking the hard soil surface crust.

2.1.2 Demerits

Due to frequent cultivation, it may cause injury to the feeding roots and the trees may be short lived or stunted in growth. Organic matter content will be completely depleted rapidly due to frequent cultivation. Frequent cultivation causes more soil erosion and loss of nutrients due to excessive leaching. Frequent cultivation may cause more soil compaction which may lead to water infiltration problems. There is more fuel and labour costs due to regular cultivation of the orchard. Unnecessary cultivation when the soil is wet, not only increases operating costs but may also cause soil damage.

2.2 Sod Culture

Sod culture serves as a basis for orchards' sustainable development by enhancing organic matter content, improving physical properties, strengthening soil water holding capacity, improving microclimate of an orchard, increasing fruit quality, and by reducing disease and pest damages. In this method, permanent cover of grass is raised in the orchard and no tillage is given. This type of orchard cultivation is followed in USA and Europe. This may be useful in slopy lands for preventing soil erosion. Sod is grown to an orchard's ground covered with grass for a long term regularly. Sod culture in orchards can be also classified into sod culture in a whole orchard and in rows. For example, sods can be grown in the whole orchard provided soils are fertile and roots grow deeply. If an orchard is infertile, with a poor drainage system, sods can be grown in rows. As for the grass, it can be annual, biennial and perennial, of which Poaceae and Fabaceae grasses are most common. This practice is followed in the orchard located on sloppy land, particularly when the gradient of the slope is greater than 10 per cent.

In this system, grasses are allowed to grow in the interspaces between the trees without tillage or mulching. Tree basins are kept free of weeds by tillage or application of herbicides. Under tropical and sub-tropical condition, sod or no-tillage in the orchard is detrimental. It negatively affects the tree growth and development. In hilly areas or where the slope of the land is elevated, trees are often grown under sod to protect the soil from erosion. Grasses are planted or the natural grasses are allowed to grow on the vacant space and the soil is not disturbed. The grass is mowed time to time and preferably falls on the ground. For tropical fruit orchards existing on the fairly levelled plain land, this practice is not suitable. The permanent grass sod between the tree rows minimizes soil erosion, increases soil aeration and permeability, and supports equipment movement through the orchard during wet weather. Intensive tillage rapidly changes the physical state of soil that has been in sod. Soil structure improves when cultivated land is put into grass. Soil aggregate distribution, stability, air permeability and hydraulic conductivity improve with time in a grass culture. According to Kay (1990), a minimum of 10 years is needed for getting maximum benefits in soil structural stability from a sod culture. In grape, though the short-term productivity is usually greater with cultivation than with sod management, however, the deleterious effects of long-term cultivation in vineyards have been recognized for many years (Wander, 1946). Long-term vineyard yield declines have been attributed to decreased organic matter and N supply and to reduced soil permeability associated with continuous cultivation. After 8 years under dry land conditions, Van Huyssteen and Weber (1980) found that grape vines growing under no tillage or mulch outperformed those under clean cultivation or sod as measured by root density, shoot growth rate, cane pruning weight, and grape yield. Sod cover has been reported to enhance P-uptake by young apple trees (Haynes 1981) and K (Bould and Jarrett 1962; Haynes 1981). The enhanced uptake of P might be attributed to a concentration effect due to lower N uptake or increased P uptake as a result of mobilization and release of P by the sod. Sod culture influences the quality of wax apple fruit, which includes the soil properties, fruit weight and sugar (Lin, 2014). Gormley *et al.* (1973) studied effect of soil management on the mineral composition of Golden Delicious and Cox's Orange Pippin apple trees and they revealed that leaf magnesium was found to be increased in Golden Delicious and Cox's Orange Pippin apple trees in association with sod cover and minimum under clean cultivation treatment. Shallow rooted grasses such as Kentucky bluegrass or annual blue grass deplete less moisture from the orchard soil profile than deep rooted sods such as clover and ryegrass. Competition of sod with fruit trees also modify the pattern of water depletion with more water being used from deeper in the profile. In summer, regular mowing of grasses is suggested to conserve moisture in soil by reducing water use by grasses. On the other hand, sod improves other physical properties as reflected by decrease in bulk density and increase in soil porosity. Due to decrease in bulk density there is increase in total pore space available for root growth and increase in water holding capacity. Such changes are likely to increase earthworm populations, reduce surface moisture runoff and erosion, and conserve moisture. Soil under permanent grass may begin to lose water 15-30 days earlier in the spring as compared to the cultivated soil. Grasses in orchard require a large proportion of water for maintaining growth, and therefore in non-irrigated orchards this competition of water with fruit trees can become critical. Potassium returns are high from sod clippings.

2.2.1 Types of Sod Culture

1. **Sod Culture Method:** Here, grasses are allowed to grow on the orchard in an unattractive manner. In this case, the grasses are neither mowed nor grazed by the animals.
2. **Sod Pasture Method:** In this method, in the orchard, both plantation of fruit tree as well as pasture ground for grazing animals are carried out. In Indian condition, this method is most common where sod culture is followed.
3. **Sod Mulch Method:** Here grasses grown in the orchard are mowed down and cut grasses are left on the land *in-situ.*
4. **Temporary Sod:** Here, the sod grasses are grown for two years, then the soil is ploughed, cultivated and re-sowing of grass seed is done.

2.2.2 Commonly Used Grasses in Sod Cover

1. Orchard grass (*Dacylis glomerata*)
2. Tall Fescues (*Festuca arundinacea*)
3. Timothy grass (*Phleum pratense*)
4. Red clover (*Trifolium pratense*)
5. White clover (*Trifolium repens*)
6. Lucerne (*Medicago sativa)*

2.2.3 Merits

1. Reduction in soil erosion and dust levels.
2. Use of low input and maintenance.
3. Suppression of weeds.
4. Sod culture could increase soil fertility, improve soil physical properties, relieve soil temperature change, increase soil microbial growth and

soil enzyme activity, improve microclimate and fruit quality, reduce physiological disease, insect pests attack (Bai *et al*.,2017).

5. Avoid bruising of falling fruits.
6. Provides opportunities for grazing of animals.

2.2.4 Demerits

1. Need for increased manuring and water application.
2. They are harmful to shallow rooted trees.
3. Decrease in Nitrogen availability to the plant roots.
4. Hence sod may be useful with deep rooted trees because soil moisture will be very low on the top layers. The roots of large trees on the other hand penetrate the soil to a greater depth and are less likely to suffer from such competition.

2.3 Sod Mulch

If the sod mulch is chosen for a period of years, always the sod grass should be mowed and left it upon the soil. The sod mulch should never be used in a young orchard. This is similar to sod culture with the only difference is that the vegetation is cut frequently and the cut material is allowed to remain on the ground. This is slightly better than sod culture, as the moisture loss is not so great as in sod culture. In both sod culture and sod mulch, more nitrogen should be applied to the fruit trees than usual application because the vegetation utilises more soil nitrogen. Tree basins are mulched with straw, instead of ploughing. The grasses are mowed down regularly and are spread to maintain the mulch in basins. This method is best suited in orchards on steep slopes to prevent soil erosion.

2.3.1 Merits

1. It has all the advantages of sod culture method.
2. It is most suitable for orchards on slopes having greater gradient.
3. It is highly effective for controlling soil erosion.
4. It hastens microbial activity, minimizes soil compaction, reduces soil erosion and makes better availability of nutrients.
5. This method can conserve moisture and control weeds in tree basins.
6. It improves fruit colouration and enhances fruit maturity.

2.3.2 Demerits

1. Under sub-tropical climatic condition, where there is prevalence of high temperature observed during summer; the sod mulch may invite fire hazard.
2. Sod mulches may attract rodents and thereby root damage.

References

Bai R, Wang Y, Zhisheng MA and Yunyun JIA (2017). Research advance on Sod Culture in Peach Orchard. *Agricultural Science & Technology*, **18** (2): 270-273, 284

Bould C and Jarrett RM (1962). The effect of cover crops and NPK fertilizers on growth, crop yield and leaf nutrient status of young dessert apple trees. *J. Hort. Sci.*, **37**: 58-82.

Chan KY, Heenan DP and Ashley R (1994). Seasonal changes in surface aggregate stability under different tillage and crops. *Soil and Tillage Res.*, **28** :301-314.

Gormley R, Robinson D and Okennedy N (1973). The effect of soil management systems on the chemical composition and quality of apples. J. *the Science of Food and Agric.*, **24**: 227-239.

Haynes RJ (1981). Some observations on the effect of grassing down, nitrogen fertilization and irrigation on the growth, leaf nutrient content and fruit quality of young Golden Delicious apple trees. *J. Sci. Food Agric.*, **32**: 1005-1013.

Hermawan B and Cameron KC (1993). Structural changes in a silt loam under long-term conventional or minimum tillage. *Soil and Tillage Res.*, **26**: 139-150.

Kay BD (1990). Rates of change of soil structure under different cropping systems. *Advances in Soil Science,* **12**: 1-52.

Kumar N (2010). *Introduction to Horticulture*. Oxford and IBH Publishing Co. Pvt. Ltd., New Delhi.

Lin YH (2014). Effects of sod culture on soil properties and fruit quality. *Kaohsiung District Agricultural Newsletter*, 89: 24-25.

Ojeniyi SO and Dexter AR (1979). Soil factors affecting the macrostructures produced by soil tillage. *Transactions American Society of Agricultural Engineers,* **22**: 339-343.

Singh J (2018). *Fundamentals of Horticulture*. Kalyani Publication, Ludhiana.

Van Huystteen L and Weber HW (1980). The effect of selected minimum and conventional tillage practices in vineyard cultivation on vine performance. *S. Afr. J. Enol. Vitic.*, **1**: 77-83.

Wander IW (1946). Physical and chemical studies of soils in north central Ohio vineyards. *Ohio Agri. Expt. Sta. Bul. 663.*

e-resources

http://ecoursesonline.iasri.res.in

Outcomes Assessment

Part A

Answer the following questions true or false.

1. Sod culture and sod mulch are synonyms. True/false
2. Clean cultivation in orchard is extensively followed in India. True/false
3. In sod pasture method, animal grazing is not allowed. True/false
4. Red clover is commonly used grass in sod cover. True/false
5. Clean cultivation does not have any beneficial effects on productivity of the orchard. True/false

Part B

Answer the following questions.

1. Differentiate between sod culture and sod mulch.
2. What is clean cultivation in orchards?
3. Define sod culture,
4. Mention one demerit of sod mulch.
5. Cut grasses are left upon the soil in sod culture/ mulch.

Part C

Write a brief essay on each of the following topics.

1. Sod culture.
2. Clean cultivation.
3. Sod mulch.

3

Herbicides

3.1 Introduction

Weed is defined as an unwanted plant growing at a place where it is not desired. According to Jethro Tull, the father of Weed Science, weed is defined as the unwanted, undesirable plant, growing out of their proper place, which interfere with the utilization of natural resources, prolific, persistent, competitive, harmful and even poisonous in nature and can grow in adverse climatic conditions. Weeds are considered major obstacle in agricultural production systems particularly in fruit crops. Weeds are a major problem to horticultural production as they take advantage of their initial slow growth rate. It has been reported that about 36-42% losses may occur due to inadequate management of weeds in apple (El-Metwally and Hafez, 2007). Fruits have a great potential but orchards are generally, infested with various types of annual, biennial and perennial weeds which compete with the fruit plants for nutrients, space, moisture, light and adversely affect growth, vigour, fruit set, flower initiation, yield, fruit quality and winter hardiness (Majek *et al.*, 1993) and thereby directly reducing the productivity of fruit trees. Weeds also provide shelter to various pathogens by becoming an alternate or collateral host of invaded crops by a number of fungal, bacterial and viral diseases. Therefore, management of weeds in all agro-ecosystems is crucial to sustain our crop productivity and to ensure the food security to the rapidly increasing population. Weed control in orchard can be usually accomplished by various methods like manual, mechanical and chemical means. Yet the conventional hand weeding is the most common method of weed control which is not only laborious but also time consuming and expensive owing to high wage rates. If fruit orchards are located in high hilly areas with small and narrow routes, it would restrict the use of machinery which is also expensive and might injure the roots and stems of the fruit trees. Gradually, manual and mechanical weed control methods are being replaced by other alternatives such as herbicides and mulches as these are easier, cheaper and less time consuming. In recent years, the use of herbicide is becoming popular among the fruit growers all over the world. It is observed to be highly effective and economical in fruit orchard besides giving the satisfactory level of weed control when applied as pre-emergent or post-emergent. Although use of various herbicides is reported to be effective in controlling weeds but the types and doses of herbicides vary from location to location depending upon the type of weed population. Therefore, the use of herbicides in fruit orchard should be done judiciously.

The chemicals employed for weed control are commonly referred to as weedicides or herbicides. Only such chemicals are used for which the standing crop has a degree of resistance or tolerance. Several types of chemicals may be used on the basis of their effect on weeds. There are two types of herbicides, one which is sprayed before the germination of weed seed, is called as 'pre-emergent' herbicide and the other which is sprayed on the weed, when they are 15-20 cm tall, is called as 'post-emergent' herbicides. Herbicides are the most important weed control tools for alleviating the infestation of weeds and getting higher yield as reported by Ashiq *et al.* (2007). Herbicidal weed control seems indispensable and has proved efficient in controlling weeds (Kahramanoglu and Uygur, 2010) and hence presently about two-third, by volume, of the pesticides used worldwide in agricultural production are herbicides. New fabrics are being developed with white colour on top (to reflect light and heat) and black on bottom to suppress weeds (Makus, 2007). Glyphosate is an effective herbicide against perennial weeds. Repeated applications of this herbicide have been found effective against many perennial weeds (MacDonald *et al.*, 2000). Barros *et al.* (2005) reported that herbicide efficiency in controlling weeds differs according to weed species. Lisek (2014) revealed that integrated weed control based on rational use of herbicides and alternative method suppressed weed competition in fruit trees. Synthetic herbicides were most effective way of controlling weeds within fruit crops but synthetic herbicides need to be used carefully so that problems with their use can be minimized. Khairnar *et al.* (2014) reported that initially after the herbicide treatment (15 days, 30 days and at harvest) microbial counts were slightly less in pre-emergence application of pendimethalin and Pendimethalin + Imazethapyr, reaching a maximum between thirty days after sowing and at harvest. They also observed that, toxic effect of herbicides normally appears immediately after application when their concentration in the soil was highest.

3.2 Losses Caused By Weeds

1. On a global scale, the potential crop yield loss without weed control was estimated as 43% (Oerke 2006). However, Rao (2000) reported the annual loss of agricultural produce due to weeds as 45%, insects as 30%, diseases as 20% and other pests as 5%.
2. They reduce the crop yield and decline the quality of produce and hence reduce the market value of the produce.
3. Weed may act as alternate hosts for many crop pests and diseases.

4. Weed control may take about 30% of total cost of expenditure thereby causing reduction in the net returns.
5. Weeds physically hinder the easy flow of irrigation water and cause its wastage by consuming the bulk of irrigation water.

3.3 Weed Management in Fruit Orchard

In cultivated crops and established orchards, the weeds can be suppressed by various methods such as chemical, mechanical, manual, biological and by mulching etc. Although the chemical weed management is most effective, it has its own constraints like the injury to non-target vegetation, crop injury, residues in soil and water, toxicity to non-target organisms. The conventional method of managing weeds through manual method is very expensive and labour intensive. Mechanical control of weeds in established orchards is rather difficult and less effective due to spreading canopies of trees, limited coverage of the implements and potential damage to root and shoots of fruit trees. Mulching or covering the soil with organic or synthetic materials has been recorded as a safe method to control weeds in comparison to herbicides application (Ramakrishna, 2006). Weed management is an integral part of an overall orchard management system. A good weed management programme should start before trees are planted. The more difficult-to-control weeds (particularly perennial species) are easier to manage before trees are planted. Weed control in orchards enhances the establishment of newly planted trees and can improve the growth and yield of established trees. Weed management strategies for higher production requires extensive knowledge of weed biology, herbicide application and calibration procedures, herbicide efficacy against target weeds and correct timing of application (Altland *et al.* 2003). The most common reasons for ineffectiveness of herbicides are improper timing, improper rates, and wrong selection of herbicide for the prevalent weed species.

3.4 Major Weeds in Fruit Orchards

A number of weeds have been reported in fruit orchards and some of the prominent weeds are enlisted as under in Table 3.1 and Table 3.2:

Table 3.1: Common weeds found in fruit Orchard

Sl. No.	Scientific name	Family
A. Grasses		
1	*Dactyloctenium aegyptium*	Poaceae
2	*Perotis indica*	Poaceae
3	*Heteropogon contortus*	Poaceae
4	*Cynodon dactylon*	Poaceae
5	*Chloris barbata*	Poaceae
B. Broad leaved		
1	*Aristolochia bracteata*	Aristolochiaceae
2	*Abutilon indicum*	Malvaceae
3	*Celosia argentea*	Amaranthaceae
4	*Hibiscus micranthus*	Malvaceae
5	*Leucas aspera*	Labiatae
6	*Leucas utricaefolia*	Labiatae
7	*Ocimum canum*	Labiatae
8	*Oldenlandia umbellate*	Rubiaceae
9	*Phyllanthus maderaspatensis*	Euphorbiaceae
10	*Trichodesma indicum*	Boraginaceae
11	*Tridax procumbens*	Asteraceae
12	*Cyanotis culculata*	Commelinaceae
13	*Tephrosia purpurea*	Leguminosae
14	*Mimosa pudica*	Mimosoideae
15	*Vicoa indica*	Asteraceae
16	*Gompherena decumbens*	Amaranthaceae
C. Sedges		
1	*Cyperus rotundus*	Cyperaceae

Source: Sivmurugan *et al.* (2011)

Table 3.2: Weed hosts for insect-pests, viral diseases and nematodes in fruit crops

Fruit Crop	Common name	Scientific name	Weed hosts
A. Insect- pests			
Citrus	Fruit sucking moth	*Othereis futlonica*	*Tinospora cordifolia, Cocculus hirsutus*
		Othereis materna	*Cocculus pendulus, Quisqualis indica, Acbysantbes* spp
Cashew nut	Bark feeder	*Indarbela tetraonis*	*Pettophorum femtgineana*
B. Viral diseases and their vectors			
Banana	Mosaic and heart rot	*Aphis qossypii, Myzus persicae*	*Achyranthes aspera, Amaranthus polygamus, A. spinosus, Datura fistula, D. metel, D. stramonium, Corchorus acutangulus, Pbysallsl floridana, Dianthus barbatus, Petunia axillaries, Xantbium strumarium, Celosia argentea, Digera arvensis, Chenopodium album, C. murale, Medicago sativa, Solatium nigrum, Euphorbia geniculate.*
	Bunchy top	*Petalonia nigronervosa*	*Colocasia sp.*
Papaya	Leaf curl, Ring spot	*Bemisia tabaci, Aphis gossypii*	*Zinnia elegans, Datura stramonium, Chenopodium quinoa*
Coconut	Root wilt, Cadang Cadang	*Strphanitis typipus, Aphis gossypii*	*Chenopodium amaranticob, Borreiria articularis, Commelina bengalensis*
C. Nematodes and related diseases			
Citrus	Slow decline	*Tilenchtdus, Semipenitrans*	*Andropogon rbizomatus, Panicum sp.*
	Lance nematode	*Haplolaimus indicus*	*Cynodon dactylon*
Banana	Rhizome root rot, Black head, Toppling, Necrotic lesion, Die back	*Radopholus similis, Helicotylenchus multicinctus*	Somc grasses, *Cynodon dactylon* and broad-leaved weeds
Strawberry	Root lesion	*Pratylenchus penetrans*	*Chenopodium album, Capsella bursa pastoris, Cirsium awense, Erigeron strigosus, Lamhm amplexicaule, Plantago major, Polygonum persicaria, Senicio vulgaris, Sonchus arvensis, Stellaria media*
Many fruits	Root lesion	*Pratylenchus* spp.	*Cynodon dactylon, tribulusi* spp. *Echinocloa crusgalli, Amaranthus* spp., *Digitaria* spp. *Sorghum halepense*

Source: (Sharma, 2006)

3.5 Methods of Weed Management in Fruit Crops

Weed control has been observed as one of the most important practice in crop production because good weed control will ensure maximum yield and high quality of farm produce. Since most horticultural crops are very slow in growth especially at the early stages of their establishment, herbicides become imperative to begin weed control early enough in order to ensure high yields and quality (Njoroge,1999). Weed management is a system approach whereby whole land use planning is done in advance to minimize the invasion of weeds in aggressive forms and give crop plants a very strong competitive advantage over the later. The systems approach is called integrated weed management (IWM). There are four methods of weed control. The first step in effective weed management is the accurate identification which in turn will help in a basic understanding of the weeds' life cycle. Correct identification can be an important step in making sure that new weeds can be eradicated before they become established.

a) **Mechanical**: In this method, weeds are controlled mainly by hand weeding, digging etc. Hand weeding is effective against annuals and biennials. In fruit nurseries, hand weeding is the most commonly and widely used method of weed control.

b) **Cultural:** Cultural methods rely upon weed control by modification of cultural practices like selection of crop, use of compost or manure, allowing the land to fallow, mulching etc.

c) **Biological:** In biological methods, there is reduction in weed infestation by direct or indirect action of biological entity such as plants, parasites, predators and pathogens. Natural enemies of plants like insects and disease-causing organisms are used. *Lantana camara* weed has been successfully controlled by torcid moth.

d) **Chemical:** In this method, chemicals are used to kill the weeds. The chemicals employed for weed control are commonly referred to as weedicides or herbicides. Only such chemicals are used for which the standing crop has a degree of resistance or tolerance. Several types of chemicals may be used on the basis of their effect on weeds. There are two types of herbicides, one which is sprayed before the germination of weed seed, is called as 'pre-emergent' herbicide and the other which is sprayed on the weed, when they are 6-8 inches tall, is called as 'post-emergent' herbicides. Annual weeds can be controlled with pre-emergence herbicides before planting an orchard. Any established annual weeds should first be controlled with post-emergence herbicides. A judicious combination of chemical and cultural methods of weed control would not only reduce the expenditure on herbicides but would benefit the crop by providing proper aeration and conservation of moisture (Prakash *et al.* 1991). Post-emergence herbicides have an advantage over pre-emergence herbicides in that post-emergence herbicides can be selected based on weed species that are present. Pre-emergence herbicides on the other hand are often applied without prior knowledge of weed species that are present in the soil, resulting in poor weed control and loss of revenue (Case and Mather, 2006). Some commonly used herbicides are listed below:

Pre-emergent herbicides: Alachlor, Atrazine, Butachlor, Diuron, Fluchloralin, Pendimethalin, Oxyfluorfen etc.

Post-emergent herbicides: Fernoxone, Gramaxone, and Glyphosate

3.5.1 Calculation of Herbicidal Doses

The containers of all commercial herbicides have label expressing the amount of active chemical contained in a particular product either in percentage of active ingredient (a.i) or acid equivalent(a.e.). All the recommendations are made on this basis. The following formula may be used to calculate the quantity of a commercial product required to give a specific dose of active ingredient.

$$Weight\ of\ herbicide\ required = \frac{Dose\ of\ a.i.\ required \times 100}{Percent\ a.i.\ in\ a\ commercial\ product}$$

Suppose, a grower wants to spray weed infested field at 1.0 Kg a.i. per hectare with a herbicide containing 50 per cent active ingredient. Then the quantity of herbicides required will be calculated as under:

$$Amount\ of\ herbicide\ required = \frac{1 \times 100}{50} = 2.0\,kg$$

3.5.2 Preparation of the Herbicide Mixture

Herbicides may be available in powder or solution form. Weighed herbicide powder may be taken in a plastic beaker and make a fine paste by adding small quantity of water. Mix this paste in the remaining water (1000 litres per hectare) with constant stirring.

3.5.3 Method of Spraying Herbicides

The prepared herbicide solution is poured into the sprayer tank using a sieve or muslin cloth to avoid clogging of the spray nozzle with dust particles and applied uniformly to the soil as pre-emergent spray or on weeds if it is a post-emergent spray. Normally, few drops of detergents like 'teepol' is added to post-emergent spray solution to increase its efficiency. Generally, 1200 litres of water per hectare is required for a post-emergent spray. For pre-emergent spray, land is thoroughly prepared without any clods followed by removal of all extraneous material like dry grass, nuts and stolons before giving the spray.

3.5.3.1 Different Methods of Herbicide Application

Several types of herbicides may be used on the basis of their effect on weeds. Broadly, they have been classified into two types viz., selective and non-selective. Selective herbicides kill special types of weeds on the other hand, non-selective herbicides kill all types of plants. These weed controlling chemicals are applied mainly through following two methods:

a) **Foliar Application:** After working the quantity of herbicide, the spray solution is prepared and applied to the weeds and other undesirable vegetation with the sprayer. These chemicals kill the weeds either by contact action or get translocated into the various parts of the plants through xylem and phloem and damage the plants. The examples of contact herbicides are Diquat, Paraquat and Nitrofen whereas herbicides like 2,4-D,2,4,5-T, Dalapon, MH etc. are translocated into the plant system and kill the plants.

b) **Soil Application:** Herbicides can also be applied through the soil before or after sowing/transplanting of the crop. Atrazine, Propazine, Nitrofen are suitable for soil application. Further, they have been named as soil fumigants or soil sterilants. The fumes of certain herbicides like Methyl bromide, Chloropicrin kill the plants and effect can persist up to one month. On the other hand, soil sterilants make the soil unable to support any vegetation. All types of plants fail to grow with the use of these sterilants, namely, Diuron, Monuron Atrazine, Simazine etc.

3.6 Weed Management in Fruit Crops

The weeds found in most of the fruit orchards like *Cynodon dactylon* (Bermuda grass) and *Cyperus rotundus* act as alternate hosts of different pests like grass hopper and *Parthenium hysterophorus* (congress grass) for mealy bugs, which attack mango, guava, pear and many other fruit trees. Several kinds of herbicides such as simazine, atrazine, terbacil, diuron, glyphosate, paraquat, bromacil etc. were found to be highly effective which could be used in destroying wide range of weeds in fruit nurseries and orchards (Taylor and Woodward, 1977; Kestov *et al.* 1983). According to Bajwa *et al.* 1976) Simazine and Atrazine produced effective results against broad leaved weeds. Excellent weed control with oxyfluorfen, atrazine and simazine has been reported by Nishimoto and Yee (1980). Glyphosate or glufosinate were more effective at controlling monocotyledonous weeds (Lee *et al.*, 1997). Rana (1998) observed that pre-emergence weedicides could control the weeds only partially, but a combination of pre-emergence and post-emergence herbicides exhibited a better weed control efficiency. Among the pre-emergence herbicides, atrazine at 2.0 kg/ha was observed to give better weed control efficiency as compared to others, whereas among the combinations, the Atrazine at 1.0 kg/ha plus glyphosate at 0.8 kg/ha was found to be the best treatment. (Table 3.3 and Table 3.4)

Mango: Three to four ploughing are enough to control weed in mango in a year. Cover cropping is another practice followed in mango orchard to suppress the growth of weeds, to bring additional income to the grower until the trees begin to bear and improve the health of trees if the intercrops grown are of right type. The recommended intercrops in mango orchards are bottle gourd, bitter gourd, onion, chillies, cowpea, black gram and green gram. Paraquat (3.0 kg a.i./ha) as post-emergence; Diuron at 6.67 - 8.9 kg/ha as pre-emergent application control weeds, effectively. Application of Atrazine or Diuron at 2.0 kg a.i./ha as preemergent followed by Paraquat at 3.0 kg a.i./ha as post-emergent have been found effective for controlling both monocot and dicot weeds in one to five years old mango orchard. Patel *et al.* (2010) claimed that pre-emergence treatment of Atrazine 2 kg a.i./ha was found superior to minimize weed population, with minimum dry weight of weeds and higher weed control efficiency in mango seedling nursery.

Banana: The major monocot weeds of banana fields are *Cyperus rotundus, Cynodon dactylon, Digitaria marginata, and Eleusine aegyptium*, and the dicot weeds are *Euphorbia* spp, *Polygonum plebejum, Portulaca oleracea, and Mimosa pudica.* Banana crop has superficial root system and thus heavy machinery for its cultivation must be avoided. The methods of weed control for banana are hand weeding, mulching, cover cropping and by application of herbicides. Inter-cropping with ginger, cowpea, colocasia and tapioca in various cultivars of banana has been reported. Herbicide has been used to weed control in banana. In banana good control of both grassy and broad leaf weeds for 4-5 months was achieved with the pre-emergent spray of diuron at 2.0 kg a.i./ha. Paraquat @ 1.8 kg a.i. /ha was used as a post emergence spray on 6-9 inches tall weeds in banana and control them for two months. Herbicide application was found to be more economical than hand weeding (6 hand weedings per year were necessary in banana crop). Pre-emergence application of Atrazine or Diuron at 1.5 to 2 kg/ha soon after planting would keep the ground weed free for 3 to 5 months. At post emergence, Paraquat or Dalapon or 2,4–D or Glyphosate can be used. (Kabeer and Nair 2009). Weed population is reduced in banana plots when *Nelsonia canescens* is used as a cover crop (Fongod *et al.* 2010).

Citrus: Grasses that cause problems to citrus orchards are generally perennials and reproduce by seeds/rhizomes, for example Bermuda and torpedo grass, or by vegetative methods, *i.e.*, by tubers or bulbs. These methods include hoeing, mechanical cultivation, mowing, slashing and burning. Mechanical cultivation is convenient in inter-row area. In citrus orchards weed control is carried out by 2 or 3 cross ploughing. While resorting to manual weeding tillage should not be deeper than 10 to 15 cm. The Spray of glyphosate 41 SL @ 1.6 litres per acre as post-emergence spray during the second fortnight of March followed by glyphosate 41 SL or gramaxone 24 WSC @ 1.2 litre per acre during second fortnight of July in 200 litres of water is effective for weed management in citrus orchards (Kaur and Rattanpal 2017). Bromacil and diuron combination at 2.0 kg a.i./ha each was found to be superior to single spray and got weed control upto 4 months. Josan *et al*. (2003) recommended glyphosate at 2.5 l/ha for effective weed control in citrus nursery.

Papaya: Being a shallow rooted plant, weeds compete with papaya plants for nutrition. Therefore, regular weeding is essential for vigorous plant growth. Weeding is usually done by hoeing or tilling. Use of herbicides like Diuron @2.0-2.5 kg a.i/ha or Glyphosate @1-2 kg a.i./ha were effective in controlling the weeds in papaya orchards. Care should be taken while spraying so as to avoid spilling of the herbicides on papaya leaves. Effective use of Paraquat and Glyphosate in papaya is dependent on avoidance of spray contact to green bark and foliage. Pre-emergence herbicide tolerance is dependent on papaya age, size and maturity, and soil type. Only one herbicide, oryzalin is shown to be tolerated by papaya immediately after transplanting. Herbicides such as Diuron and Oxyfluorfen with a broader spectrum of weed control generally injure young papaya, however they can be effectively used if the initial application of these herbicides is delayed until papaya attains certain size or maturity indices (Nishimoto, 1997).

Grape: Atrazine or Simazine or Diuron at 2.0 kg a.i./ha as preemergent followed six months by two post-emergent sprays of Paraquat at 2.0 kg a.i./ha at three monthly intervals for grape cv. Banglore Blue. Oxyflurofen at 1.0 kg a.i./ha as pre-emergent spray and Glyphosate at 3.0 kg a.i./ha as post-emergent spray proved very effective for controlling weeds for four and six months respectively in grape. (Rana and Rana, 2016).

Guava: The problematic weeds of guava fields are *Cynodon dactylon, Cyperus rotundus, Bidens pilosa, Tridax procumbens, Acanthospermum hispidum and Lagasca mollis. Cynodon dactylon* caused 50 per cent reduction in plant height of guava. The weed in guava orchard can be managed by following hand weeding, mulching and by application of 2,4-D alone at 3.0 kg/ha or 2,4-D (2.0 kg/ha) in combination with gramoxone (1.5 l/ha). Manual weeding and herbicide combination of Glyphosate (0.5 kg a.i./ha) + 2, 4-D (1 kg a.i./ ha) were valuable practices in guava (cv. L-49) weed management (Maji *et al.* 2008).

Pineapple: The common weeds in pineapple plantations are Hariyali and nutgrass. Weeding is important from the economic point of view. Hand weeding especially in closely spaced crop is cumbersome and uneconomic. Therefore, chemical control of weed is advisable. Pineapple Research Station, Vellanikkara, Thrissur (Kerala), recommended application of Diuron at 3 kg/ha or combination of Diuron (1.5 kg/ha) + Bromacil (2.0 kg /ha) as pre-emergence spray. Pre-emergence herbicides like Simazine or Atrazine or Diuron @ 3 to 4 kg/ha application at 4-6 weeks before planting and ploughing the field at 2 weeks before the leaves of the crop began to curled down to prevent weed establishment. Weed growth emerging later is treated with post emergence herbicides such as Glyphosate @ 0.5 kg/ha or Dalapan @ 3 to 4 kg/ ha or 2,4 - D @ 0.5 to 1 kg/ha depending upon the predominant weed species. Dalapan spray drift can cause chlorosis on outer leaves and necrosis on leaf tips when pineapple plants are young. It should be applied carefully any time from 6 months.

Apple: Apple (*Malus domestica*) grows mostly in the temperate regions of Jammu & Kashmir, Himachal Pradesh and Uttrakhand and to small extent in Nilgiris Hills of Tamil Nadu. *Rosa moschata*, *Rosa eglantaria*, *Rubus* sp. and *Berberis* sp. are some of the predominant weeds of apple orchards. The weeds can be controlled by applying mulching with black polythene and oat leaves in apple basins, growing intercrops (mustard, wheat, oat, barley). Chemically, weeds can be managed with the spray of gramoxone at 500 ppm + Ansar 529 at 1000 ppm when applied as post-emergent spray. Gramoxone (3.75 kg/ha) was found to be the most efficient post-emergent herbicides out of a number of herbicides tested in apple orchards. An application of Aminotriazole (5.0 kg/ha) and Simazine (5.0 kg/ha) as pre-emergent spray resulted in the elimination of broad leaf weeds and increases in the density of grasses (Rana and Rana, 2016). Application of paddy straw mulch followed by glyphosate was appreciably effective in improving growth parameters as well as leaf and soil nutrient status. In general, to meet the multiple objectives, paddy straw mulch followed by glyphosate provided good weed control in apple orchard and its adoption is beneficial to crop and the soil to represent a good choice in orchard floor management system (Hussain *et al.*, 2018).

Plum: Gautam (1980) while conducting an experiment on effect of long-term use of several herbicides on yield, quality and storage of Santa Rosa plum reported that all herbicide treatments significantly reduced fruit yield compared with the hand-weeded control. He reported that treatments of simazine or diuron each at 6 kg/ha and atrazine at 6 kg/ha or terbacil at 3 kg/ha reduced fruit acidity and the TSS: acid ratio but had no effect on plum storability at room temperature. Significant improvement in fruit size, fruit weight, juice percentage, total soluble solids, acidity and total sugar content under cumulative doses of 6.8 and 4.4 litres/ha glyphosate was reported by Minhas *et al.* (1994).

Table 3.3: Herbicides for weed control in some fruit crops

Fruit crops	Application	Dose
Mango	Paraquat	3.0 Kg a.i./ha
	Atrazine or Diuron	2.0 Kg a.i./ha
Banana	Paraquat	1.8 Kg a.i./ha
Grape	Atrazine or Simazine	2.0 Kg a.i./ha
	Paraquat	2.0 Kg a.i./ha
	Glyphosate	3.0 Kg a.i./ha
Papaya	Alachoor or Butachlor	2.0 Kg a.i./ha
Sweet Orange	Bromacil + Diuron	2.0 Kg a.i./ha each
Mandarin	Monouron or Diuron	4.0 Kg a.i./ha
Guava	2,4-D	3.0 Kg a.i./ha
Sapota	Bromacil + Diuron	2.0 Kg a.i./ha each
Pineapple	Bromacil + Diuron	2.0 Kg a.i./ha each
	Glyphosate	1.5 Kg a.i./ha
Ber	Oxyfluorofon	0.6 Kg a.i./ha
	Oxadiazon	1.5 Kg a.i./ha
Pomegranate	Simazine + Oryzalin	2.0 Kg a.i./ha each
Anola	Aminotriazole	4.0 Kg a.i./ha
	Paraquat or Diquat	0.5-1.00 Kg a.i./ha

(Maji *et al.* 2004 and Makhija and Singh, 2011)

Table 3.4: Herbicide application in horticultural crops

Crop	Chemical Name	Dose/ha	Trade Name	Dose /ha	Time of Application
Cole crops	Pendimethalin + Hand weeding	1.0 kg	Stomp 30% EC	3.3 L	3 DAT + 40 DAT
	Fluchloralin + Hand weeding	1.0 kg	Basalin 45% EC	2.2 L	3 DAT + 40 DAT
Carrot	Fluchloralin	1.0 kg	Basalin 45% EC	2.2 L	3 DAT
	Pretilachlor	0.50 kg	Rift 50% EC	1.0 L	3 DAT
Beans	Fluchloralin + Hand weeding	1.0 kg	Basalin 45% EC	2.2 L	3 DAT + 30 DAT
Radish	Fluchloralin + Hand weeding	1.0 kg	Basalin 45% EC	2.2 L	3 DAT + 30 DAT

Crop	Chemical Name	Dose/ha	Trade Name	Dose /ha	Time of Application
	Pendimethalin + Hand weeding	1.0 kg	Stomp 30% EC	3.3 L	3 DAT + 30 DAT
	Alachlor + Hand weeding	1.5 kg	Lasso 50% EC	3.0 L	3 DAT + 30 DAT
	Metalachlor	1.0 kg	Dual 50% EC	2.0 L	3 DAT + 30 DAT
Peas	Alachlor	1.0 kg	Lasso 50% EC	2.0 L	3 DAT
Cucurbits	Oxyfluorfen	250 g	Goal 23.5% EC	1 L	3 DAS
Cumin/ Fenugreek/ Coriander	Pendimethalin + Hand weeding	1.0 kg	Stomp 30% EC	3.3 L	3 DAS + 40 DAS
	Fluchloralin	1.0 kg	Basalin 45% EC	2.2 L	Preplant incorporation
Turmeric	Pendimethalin + Hand weeding	1.0 kg	Stomp 30% EC	3.3 L	3 DAS + 45 DAS
	Alachlor + Hand weeding	1.0 kg	Lasso 50% EC	2.0 L	3 DAS + 45 DAS
Tea	Oxyfluorfen	400 g	Goal 23.5% EC	1.7 L	3 DAS
Coconut	Paraquat	1.0 kg	Gramaxone 24 % SL	4.0 L	2-3 leaf stage of weeds
Problematic weed management					
Cynodon dactylon & *Cyperus rotundus*	Glyphosate+ AGF activator	4.1 g + 2 ml/lit of water	Roundup 41% SL	10 ml (Commercial product)	pre flowering stage
Under non crop situation,	Glyphosate + Ammonium sulphate	2.05 kg + 2 %	Roundup 41% SL	5.0 L + 2 %	pre flowering stage
Parasitic weed management					
Striga management in Sugarcane	Atrazine + Hand weeding + earthing up combined + 2,4-D sodium salt + urea + trash mulch	1.0 kg+ 5 g/litre (0.5%) + urea 20 g/ litre (2%)	Atratop 50% WP + Fernoxone 80% WP	2.0 kg + 5 g/litre (Commercial product)	3 DAP + 45 DAP + 60 DAP + 90 DAP + 150 DAP
Cuscuta	Hand removal & Paraquat	0.80 kg	Gramoxone 24% SL	3.3 L	20 DAS
	Pendiemthalin	1.0 kg	Stomp 30% EC	3.3 L	3 DAT

Crop	Chemical Name	Dose/ha	Trade Name	Dose /ha	Time of Application
Solanum elaeagnifolium	Glyphosate alone or in combination with 2-4, D Na salt	1.75 kg +2.0 kg/ha	Roundup 41% SL + Fernoxone 80% WP	4 L + 2.5 kg	25 DAT
	Glyphosate + 2,4-D + soap solution	4.1 g + 6 g + 2 ml/lit of water	Roundup 41% SL + Fernoxone 80% WP	10 ml + 6 g (Commercial product) + 2 ml/lit of water	25 DAT
Water hyacinth	Glyphosate + Ammonium sulphate + surfactant	4.1 g + 20 g + 1 ml/ lit of water	Roundup 41% SL	10 ml + 20 g (Commercial product) + 2 ml/lit of water	25 DAT

Source: http://www.agritech.tnau.ac.in/agriculture/agri weedmgt horticrops

3.7 Herbicide Resistance

Herbicide resistance is the inherited ability of weeds to survive and grow at herbicide dosages many times greater than what is usually needed for control of that species. The continuous use of herbicides might lead to the development and evolution of resistant biotypes. Herbicides are considered excellent tools within a weed management strategy in many cropping systems; however, misuse of this technology can lead to problems such as residual carry-over, cropping restrictions, groundwater contamination and the development of genetically-based herbicide resistance (Booth *et al.*, 2003).The potential risk for the development of herbicide resistance is greatest when the same herbicide is used repeatedly, as is often done in orchards. To prevent the development of herbicide resistance, a variety of weed-control strategies should be adopted including cultural practices and alternating herbicides with different modes of action. Failure to do this can result in the rapid loss of herbicides as an effective pest management tool, although cultivation remains an option.

3.8 Precautions for Safe use of Herbicides

1. Herbicides should be sprayed at recommended concentration and time.
2. The sprayer must be cleaned with warm water every time after use. If possible, sprayer used for herbicides should not be used for any other kind of spray.
3. Inhaling of vapours, dust or spray mist should be avoided. Irrigate at regular intervals soon after spray of pre-emergent herbicides like atrazine, diuron, etc. Irrigation should be done at regular intervals which is

necessary to carry the herbicides to the lower layers of soil where weed seeds and nuts are germinating.

4. The soil should not be disturbed at least for a month in case of pre-emergent spray.
5. Post-emergent herbicides should not be sprayed on crop plants, as it is contact in nature and kills those parts on which it falls.
6. These herbicides must not be sprayed on a rainy day. At least a gap of 8-10 hours is required between one spray and rain for effective control of weeds.

References

Altland JE, Wehtje G and Gilliam CH (2003). Weed control in field nurseries. *Hort Tech.*,**13**: 9-14.

Ashiq M, Sattar A, Ahmed N and Muhammad N (2007). Role of herbicides in crop production. Unique enterprises 17-A, Gulberg colony, Faisalabad, Pakistan.

Bajwa GS and Singh DB (1992). Note on weed control in Patharnakh nursery. *Indian J. Hort.*, **49** (1): 58-59

Barros JFC, Basch G and de Carvalho M (2005). Effect of reduced doses of a post emergence graminicide mixture to control *Lolium rigidum* in winter wheat under direct drilling in Mediterranean environment. *Crop Protection,* **24**: 880-887.

Booth BC, Murphy SD and Swanton CJ (2003). *Weed Ecology in Natural and Agricultural Systems*, CABI Publishing, Oxon p. 303.

Case L and Mather H (2006). Field evaluation of herbicide treated mulches. *In: Proceeding Southern Nursery Association Research Conference*, **51**: 402

El-Metwally IM and Hafez OM (2007). *Arab Univ. J. Agril. Sci.*, Ain Shams University, Cairo, **15** (1): 157-166.

Fongod AGN, Focho DA, Mih AM, Fonge BA and Lang PS (2010). Weed management in banana production: The use of *Nesonia canescens* Spreng as a non-leguminous cover crop. African J. Env. Sci. and Tech., **4**:167-73.

Gautam DR (1980). Effect of long term use of several herbicides on yield, quality and storage of Santa Rosa plum. *Haryana J. Hort. Sci.*, **9** (3/4): 122-124

Hussain S, Bashir D, Sharma MK and Jamwal M (2018). Effect of Orchard Floor Management Practices on Fruit Production. *Int. J. Curr. Microbiol. App. Sci.,* **7** (7): 3627-3670.

Josan JS, Sharma JN and Monga PK (2003). Effect of various herbicides on weed population in citrus nursery. *Ind. J. Hort.*, **60** (1): 45-48

Lisek J (2014). Possibilities and limitation of weed management in fruit crops of temperate climate zone. *J. Plant Protection Res.*, **54** (4): 318-326.

Kabeer KAA and Nair VJ (2009). *Flora of Tamil Nadu- Grasses*. Botanical Survey of India, Kolkatta.

Kahramanoglu L and Uygur FN (2010). The effects of reduced doses and application timing of metribuzin on redroot pigweed (*Amaranthus retroflexus* L.) and wild mustard (*Sinapis arvensis* L.). *Turkish J. Agric. and Forestry*, **34**: 467-474.

Kaur N and Rattanpal HS (2017). Weed management in citrus orchards. *Rashtriya Krishi*,**12** (1) :122.

Khairnar CB, Goud VV and Sethi HN (2014). Pre- and post-emergence herbicides for weed management in mungbean, *Indian J. Weed Sci.*, **46** (4): 392-395.

Keshtov Yu, Alekseea A and Fedorov YuV (1983). Chemical control of weeds in fruit nurseries. *Zaschita Rastenil*, **4**: 29

Makus D (2007). Use of fabric and plastic barriers to control weeds in blackberries. *J. Subtropical Plant Sci.*, **59**: 95-103.

Majek BA, Neary PE and Polk DF (1993). Smooth pigweed interference in newly planted peach trees. *J. Production Agric.*, **6:** 244-246.

Makhija M and Singh S (2011). Weed management studies in aonla. *Haryana J Hort. Sci.* **39**:185-87.

Minhas PPS, Singh D, Sandhu, A.S. and Mann, S. S. (1994). Chemical weed control in pear orchards of Northern India. *Acta Horticulturae,* **367**: 331-341

Nishimoto RK and Yee WYJ (1980). A guide to chemical weed control in tropical and subtropical fruits and nut crops in Hawaii. Circular cooperative Extension Service, University of Hawaii, 431: 11

Nishimoto RK (1997). Herbicide options for weed control in papaya. *Integrated Pest Management Reviews*, **2**:109–111.

Njoroge JM (1999). *17th East African Biennial Weed Science Conference proceedings*, pp: 65-71.

Oerke EC (2006). Crop losses to pests. *Journal of Agricultural Science* **144** (1): 31-43.

Prakash V, Prasad K and Singh P (1991). Chemical weed control in soybean. *Indian J. Weed Sci.*, **23** (1-2): 29-31.

Ramakrishna A, Tam HM, Wani SP and Long TD (2006). Effect of mulch on soil temperature, moisture, weeds infestation and yield of groundnut in northern Vietnam. *Field Crops Res.*, **95** (2-3): 115-125.

Rana MR (1998). Effect of certain nursery management practices on growth of nursery plants of apple. *M.Sc. Thesis, Dr. Y. S. Parmar University of Horticulture and Forestry,* Solan, H P India.

Rana SS and Rana MC (2016). Principles and Practices of Weed Management. Department of Agronomy, College of Agriculture, CSK Himachal Pradesh Krishi Vishvavidyalaya, Palampur,138pages. (DOI:10.13140/RG.2.2.33785.47207).
www.hillagic.ac.in /edu/coa/agronomy/lect.)

Rao VS (2000). *Principles of Weed Science.* 2nd edition. New York: Science Publishers, 526 p.

Sharma RR (2006). *Fruit Production Problems and Solutions*. International book Distributing Co., Lucknow, India.

Sivamurugan AP, Nanthakumar S, Sganthi A and Davidson SJ (2011). Weed management in fruit crops. *Rashtriya Krishi*, **6** (2): 52-52.

Taylor JC and Woodward RS (1977). Roundup tested for weed Woodward control in young peach plantings. *Lowsilana Agric.*, **20**: 10-11.

e-resource

http://www.agritech.tnau.ac.in/agriculture/agri_weedmgt_horticrops

Outcomes Assessment

Part A

Answer the following questions true or false.

1. Weed is an unwanted plant growing at a place where it is not desired. Yes/No

2. Herbicide resistance is not an inherited ability of weeds. Yes/No
3. Weeds are a minor problem to horticultural production. Yes/No
4. Alachlor and Atrazine are Pre-emergent herbicides. Yes/No
5. Herbicides must not be sprayed on a rainy day. Yes/No

Part B

Answer the following questions.

1. Define weeds.
2. What is cultural method of weed management?
3. Who is the father of weed science?
4. Orchards are infested with various types of —, — and perennial weeds.
5. Differentiate between pre-emergent and post-emergent herbicides

Part C

Write a brief note on each of the following.

1. Discuss weed management in guava.
2. Methods of weed management in fruit crops.
3. Precautions for safe use of herbicides.

4

Inorganic and Organic Mulches

4.1 Introduction

Fruit tree plantations are increasing gradually, sometimes at the expense of marginal lands. Consequently, limiting soils are now being used for crop production. Such soils are considerably constrained in terms of infiltration, retention, and transmission of water, and also regarding the salinity and/ or nutrient content. A number of agronomical techniques can be used to improve soil productivity for agricultural purposes; however, many of them are ineffective or unsustainable. Mulching techniques have emerged in recent years to address these soil constraints yielding positive results in a variety of scenarios. Zribi *et al.* (2015) reported the positive effects of mulches and organic amendments on soil physical properties such as increased water retention and decreased soil evaporation.

Mulching is the process of covering soil around plant root area with some extraneous matter of organic or inorganic in nature with a view to insulate the plant and its roots from the effects of extreme temperature fluctuations and make conditions more favourable for growing, as well as avoid weed germination and to conserve the available soil moisture.

A wide variety of materials can be used as mulches, which are broadly classified as organic and inorganic mulches. All the organic residues, *viz*., crop residue, leaves, organic wastes *viz*. farm yard manure, poultry manure, spent mushroom compost, straw, saw dust, pine needle/bamboo matting, tree bark, grass clippings, leaves, hay, kitchen scraps, comfrey, dried sewage sludge (SS), wood chips, shredded newspaper, cardboard, wool, animal manure, cocoa bean hulls, rice hulls, corn cobs or stalks, cotton seed hulls, peat moss, vermiculite or perlite, nut shells, hydro mulching etc. and inorganic mulches, *viz*., black polythene, transparent polythene, rock, stone, rubber mulch, plastic thick films, landscape fabric and marble chips etc. can be used as mulches. In some arid fruit growing areas, pebbles can be used in the event of non-availability of organic material. These mulches act as a surface barrier against evaporation, conserve soil moisture, check weed growth and enrich organic matter to the soil. However, mulches should be removed at the onset of monsoon rain.

Mulches make unfavourable conditions for seed germination of weeds and provide a physical barrier for emerging weeds. A good mulch layer saves many hours of manual weeding. Thick layer of organic mulch material is very effective in preventing the number of annual weeds in the orchard, since they have difficulty to penetrate such a layer. Mulches maintain uniform moisture conditions in the soil. Moisture loss through evaporation is reduced and soil erosion is decreased in rainy season. Slow and steady water infiltration takes place in the soil and crusting is prevented. Mulches reduce the force of raindrops, thereby protecting the structure of the soil at the surface. Mulches may serve as vapour barriers, thus reducing evaporation of soil moisture. Water fall on bare soil surface causes splashing of soil onto the fruit. Mulches keep fruits cleaner and help in reducing the spread of disease.

Organic as well as living mulches can be effective in removing heavy metals from soils. Contaminants such as lead and cadmium can be removed from the soil solution by mulched leaves of eucalyptus (*Eucalyptus* spp.), pine, poplar (*Populus* spp.) and *Thuja* spp. (Tyagi *et al.* 2018). Mulches reduce evaporation, more water is left in the soil and salts are diluted as well as salt accumulation can be avoided by mulching to prevent upward movement of saline water from deeper layers (FAO, 1990). Furthermore, organic mulches can reduce the effect of salt toxicity on plant growth (Landis 1988) or actively accelerate soil desalinization (Dong *et al.* 1996). Organic mulches can also help degrade pesticides and other contaminants (Gan *et al.* 2003), presumably by providing increasing microbial populations that degrade pesticides. Whereas, plastic mulch promotes earliness by capturing heat, which increases soil temperatures and accelerates growth. It also increases the effectiveness of soil fumigant chemicals. Because of the impervious nature of the plastic mulch, it acts as a barrier to gas escape and keeps gaseous fumigants in the soil.

4.2 Advantages

1. Conserves soil moisture by reducing evaporative losses from soil surface.
2. Organic mulches add organic matter and nutrients to the soil upon decomposition thus, save fertilizer requirements.
3. Improves the soil structure, nutrient availability and micro flora.
4. Controls soil erosion and improve the water-holding capacity of the soil.
5. Improves water infiltration and controls weed growth due to etiolation.
6. Reduces soil temperature fluctuation.
7. Mulch acts as a cushion for dropping fruits and thus avoiding physical injury.

8. Growth, production, colour and quality of fruits are improved.
9. Organic mulches also improve root growth.
10. Organic mulch provides an ideal environment for earthworms and other beneficial soil organisms.
11. Organic mulches are more useful for warmer climates, i.e., valley areas and low and mid-hills areas of Himalayan region, as these keep the soil temperature cooler by 2-3°C in summer.

4.3 Disadvantages

1. Cost of mulch is high.
2. Dry grass, hay or pine needle mulches are a fire hazard.
3. Mulch may harbour insects, diseases and rodents.
4. Inorganic mulches do not decompose in the soil, so to be removed from the field at ending of fruiting in seasonal crops.
5. Rainfall has little effect on soil moisture in plasticulture because plastic does not allow for water infiltration.
6. Causes environmental pollution.
7. Toxic to livestock

Different types of mulch materials have some special advantages and disadvantages; details are presented in Table 4.1

Table 4.1: Advantages and disadvantages of different mulch materials

Mulch material	Advantage	Disadvantage
Organic Mulch Materials		
Bark (hard wood/pine)	Slow to decompose	May hinder water penetration
Evergreen Tree Trimmings	Readily available; inexpensive or free.	Need to clear out of beds in spring.
Sawdust	Good soil conditioner; decomposes quickly; slightly acidic; free.	May deplete nitrogen in the soil.
Pine Needles	Slow to decompose; slightly acidic; suppresses weeds; light weight; free of weed seeds; does not pack down; reusable.	Requires gloves to handle; flammable when dry.
Straw	Good insulator; light weight; readily available; lightens soils when tilled in; useful when used around vegetables and as a cover for seeded lawns.	Flammable, may deplete nitrogen

Mulch material	Advantage	Disadvantage
Grass Clippings	Quickly decomposes; best if layered thinly, readily available; best to compost or age before using.	Depletes nitrogen at the soil surface; if applied too thickly, can dry and become a barrier to water, may become soggy and pack down; can carry weed seed; could contain herbicides.
Leaves	Good insulator; easy to transport; readily available; best if shredded; contain trace minerals; will work into soil to supply good organic matter; apply after soil has warmed.	Can become soggy and pack down restricting water flow.
Newspaper	Good barrier to weed seed germination; keep wet or cover with bark or other mulch; best used between vegetable rows; readily available.	Coloured newsprint may contain lead; can blow away if not covered; unsightly in landscape setting.
Cocoa Bean Hulls	Attractive and fragrant; good soil conditioner; contains about 2 percent nitrogen.	May be hard to obtain; expensive.
Rice Hulls	Slow to decompose; absorbent; good soil conditioner; best if composted first.	Requires additional mulch to hold it in place; may pack down and retard water penetration; may contain weed seeds.
Corn Cobs or Stalks	Moderately available; good weed control; supplies some nutrients.	May heat up and compost in place.
Cotton seed Hulls	Adds nutrients to soil like cotton seed meal.	Can be blown about; needs to be topped with a layer of heavier mulching material; not readily available.
Peat Moss	Clean and free of weed seeds; improves water retention and soil tilth when tilled in; readily available; decomposes slowly.	Water penetration hindered if applied too thickly.
Vermiculite or Perlite	Sterile and free of weed seed; readily available.	Expensive, very light and may scatter in the wind or be washed away; can cake to inhibit water penetration
Nut Shells	Slow to decompose; attractive; good soil conditioner; adds nutrients to the soil.	Sharp edges make it hard to work with; may not be readily available; expensive.
Inorganic Mulch Materials		
Rock	Permanent mulch; inexpensive; can be attractive.	Does not control weeds

Mulch material	Advantage	Disadvantage
Black/Clear Plastic	Retains soil moisture; black provides effective weed control; warms soil quickly in spring; readily available and fairly easy to apply; best for vegetables but can be used in and around landscape plantings.	Non-permeable to air/water; can suffocate shallow-rooted plants; non-degradable; weeds grow under clear plastic; adds no nutrients; should be covered with other mulch materials to prevent UV radiation damage.
Porous Plastic	Allows water and air to move through; same as other plastics.	Degraded with exposure to the sun; should be top mulched to protect.
Landscape Fabric	Allows water and air to move through; durable; suppresses most weeds; good for landscape and perennial beds.	Some weeds can get started in fabric; should be covered by top mulch.
Gravel or Marble Chips	Permanent mulch; inexpensive; can be attractive.	Absorbent, adds no organic matter to soil; can become hot and affect root system of shallow-rooted plants; cannot cultivate soil.

Source: Tyagi *et al.*, 2018

4.4 Organic Mulches

The mulches that add nutrients and humus to the soil as they decompose is known as organic mulches. They improve the soil tilth and moisture holding capacity. Some mulches are tilled into the soil before planting a new crop and therefore may have an effect upon soil fertility and soil property. In short term, mulches may decrease nitrogen available for a given crop. A material that has a high carbon content and is very low in nitrogen and other nutrients may actually "bind" or immobilize plant available nitrogen temporarily. This occurs because soil microorganisms use available nitrogen to metabolize and decay the organic material. The immobilized organic nitrogen can be made available (mineralized) later as the organic matter continues to decompose. The increased organic matter content due to mulches increases soil aggregation, infiltration, and water-holding capacity. In poor soils, addition of organic material improves water retention and reduces water losses through deep percolation.

4.5 When and how Much to Apply Mulch

Time of application depends on what is expected to be achieved by mulching. Mulching by providing an insulating barrier between the soil and the air, moderate the soil temperature. This means a mulched (organic) soil in summer will be cooler than the adjacent non-mulched soil; while in the winter, the temperature of mulched soil may not drop as deeply as otherwise it would have been. In winter mulches can be applied in late after the temperature falls below freezing point but before the coldest temperature arrives. Applying mulches before the coldest temperature may attracts rodents looking for a warm over-wintering site. Delayed applications of mulch should prevent this problem as, hopefully, the creatures would already have found some other place to nest. One of the benefits from winter mulching is the reduction in the freezing and thawing of the soil in the late winter and early spring. These repeating cycles of freezing in night and then thawing in the warmth of the sun causes many small or shallow rooted trees (e.g. young trees) to be heaved out of the soil. This leaves their root systems exposed and results in injury or death. Mulching also prevent the rapid fluctuation in soil temperature.

Most of the organic mulches should be applied after plants are well established. In developing plants, organic mulch can be applied but problem of termite should be taken care of. Remove all weeds before spreading the mulch. Apply mulches when there is reasonably good soil moisture and before the weather turns hot. Infiltration of rain water will be slowed somewhat by mulch, so it is best not to place the mulch on soil which is dry. Water thoroughly or wait for a good soaking rainfall before applying any mulch. Application of organic mulch in plant basin can delay the termite infestation to tree trunk for some days as termites feed on mulch material and do not move to tree. However, it aggravates the infestation in subsequent days when mulch is digested and no additional mulch is replaced. To keep these termites in check, tree trunk should be pasted with chlorpyriphos 0.02% + coal tar paste and mulch should be applied one foot apart from the tree trunk. This practice delays the termite infestation and restricts the soil mounding on tree trunk. Around 10 cm thick hay, straw or pine needles are spread over the tree basin areas in spring immediately after completing the tillage operation and applying second dose of nitrogenous fertilizers in the basins. However, about 10-15 cm area away from the tree trunk should be kept open, in order to protect the trunk from the damage of any harbouring insect-pests, rats etc.

Among the inorganic mulches black polythene and similar materials should be spread on land that has been completely prepared for planting and has a high moisture level. Use of black polythene sheet of 200-micron thickness around the tree basins or in strips is common, keeping sufficient vacant space for working between the rows. Edge of the mulch should be buried with soil to prevent it from blowing away due to air.

4.6 Types of Organic Mulch

1. **Bark (Hardwood):** Shredded hardwood is a byproduct of the lumber and paper industries and varies in size from shredded chips to larger nuggets. This type of mulch is generally used around trees and shrubs as well as in perennial beds. Shredded bark is available in natural or dyed varieties. Dyed varieties are often a mix of hardwood or recycled wood waste containing artificial dyes.
2. **Bark (Softwood):** Softwood bark is also a byproduct of the lumber and paper industries. Pine bark is a common example and is frequently used under large trees and shrubs. Pine bark is slightly acidic in nature, takes time to decompose, and does not need replacing as often as other types of organic mulch.
3. **Municipal tree waste:** City or village arborists and utility companies often make mulch available to home owners at no charge. Normally this much is made up of larger chunks of wood and is not aged. The fresh material will utilize larger amounts of nitrogen in the soil as it decomposes. This type of mulch is especially useful for creating pathways.
4. **Cocoa bean hulls/cocoa bean mulch:** Cocoa hulls are byproducts of the chocolate industry. The hulls are light weight, easy to handle, appropriate for all planting areas and smell pleasantly. Hulls should be applied not more than 1 inch deep and should be lightly watered to hold them in place. They decompose quickly and require annual application. Consider choosing another mulch if pets frequently visiting to the area as chocolate byproducts may be lethal to animals if consumed.
5. **Leaf mulch:** Leaf mulch can be created at home by composting shredded leaves. Leaf mulch can be used in all garden beds. Leaves that may be infected with scab, leaf spot, or anthracnose should be disposed of instead of composted for mulch.
6. **Grass clippings:** Grass clippings may be spread in thin layers across vegetable and perennial beds and turned in at the end of the growing season. Allow each layer to dry before adding additional layers. Do not apply in thick layers as clippings will mat. Do not use clippings from lawns that have been treated with herbicides or insecticides. Also, do not apply grass clippings that have turned to seed to prevent undesirable turf grass from growing in garden beds.
7. **Composted animal manure:** Well composted animal manure can be used as a mulch or soil amendment. Composted animal manure is an excellent choice for new planting beds as it improves soil quality and adds nutrients. Fresh manure should not be used in garden beds because

it can burn plant roots. Manure should be well composted prior to using as mulch (as temperatures reaching 54 to 60 ^{0}C) for at least one week and composted for four to six months longer to eliminate most potential disease organisms. Dog, cat and pig manure should never be added to vegetable gardens.

8. **Newspaper:** Layers of black and white newspaper can be used to suppress weeds. Apply two to three layers at a time and cover with an organic material such as leaf mulch or grass clippings to hold it in place. News print will eventually decompose and can be incorporated into the soil.
9. **Hydro-mulching:** Hydro-mulching includes wood cellulose fibre mulch (Acquaah 2009). Hydroseeding (or hydraulic mulch seeding, hydromulching, hydroseeding) is a planting process that uses slurry of seed and mulch. It is often used as an erosion control technique on construction sites, as an alternative to the traditional process of broadcasting or sowing dry seed (Tyagi *et al.*2018).
10. **Living Mulch:** The most important attributes required for species used as living mulches are quick emergence and soil covering, short height, low water and nutrients demands. In most cases the legume crops or cereals, especially rye (*Secale cereale* L.) are used for this purpose. The living mulch not only provides for the effective management of weeds but also for the decrease of insect pest pressure resulting in lower requirement for pesticide use (Tyagi *et al.*2018).
11. **Sod:** In this method, permanent cover of grass is raised in the orchard and no tillage is given. This type of orchard cultivation is followed in USA and Europe.
12. **Sod mulch:** This is similar to sod with the only difference is that the vegetation is cut frequently and the cut material is allowed to remain on the ground.

Table 4. 2: Thickness of the organic mulches to be applied around the tree basin

Organic mulch material	Thickness of mulching
Compost	3-4 inches
Green leaves	3-4 inches
Dry leaves	6 inches
Grass clippings	2-3 inches
Bark	2-4 inches
Wood chips	2-4 inches
Newspaper	¼ inch
Onion/garlic scale	2-3 inches

4.7 Inorganic Mulches

Inorganic mulches are usually preferred in fruit orchards located in the high hill areas having cooler climate. It also helps in raising the soil temperature in summer, required for greater nutrient absorption and roots metabolic activities. Now, different inorganic mulches are used everywhere as per requirement. However, inorganic mulches lack soil improving properties of organic mulches.

Plastic mulch application is effective in increasing soil temperature, conserving soil moisture and weed control (Lamont 2005). The most popular plastic mulch worldwide is black, though white-on-black and clear mulches are also used (Schales 1990). Ham *et al.* (1993) reported that white-on-black and silver mulches reflect 48% and 39% of short-wave radiation, respectively.

A wide range of plastic films based on different types of polymers have been evaluated for mulching. Among them, LDPE, HDPE and flexible PVC have been used. There are some technical performance differences among them, however it is minor in nature. Owing to its greater permeability to long wave radiation which can increase the temperature around plants during the night times, polyethylene is preferred. Today the vast majority of plastic mulch is based on LLDPE because it is more economic in use.

4.8 Types of Inorganic Mulch

1. **Plastic film (polyethylene):** Plastic film is impermeable; water and nutrients cannot pass through. Plastic is best used along rows of vegetables to warm the soil in spring and used in fruit orchards to prevent weed germination and conserve moisture. It is not the best choice for long-term use. Plastic film deteriorates with exposure to sunlight and is usually used for only one season. Opaque sheets minimize weed problems, since weeds receive partial light while germinated seeds receive full light. Solarization techniques use clear plastic mulch to heat soil to temperatures high enough to kill soil borne pests. Generally, researchers using black plastic instead of bare soil have recorded higher yield in many crops. Different types of plastic mulches are available for different purpose which are described below:

   a. **Black polythene mulch:** The black plastic film does not allow sunlight to pass through onto the soil. Thus, photosynthesis does not take place in soil in absence of sunlight below the black film. Hence, it arrests weed growth completely. The black plastic mulch is helpful in conserving moisture and controlling weed growth. However, it may increase the soil temperature. It has been reported that during hot climatic conditions, using plastic nets and non-woven fabrics (layer of polyester or poly-vinylalcohol) in place of black mulch film is also helpful in increasing yield marginally of vegetables, especially that of leafy vegetables (Lyengar et al., 2011).

b. **Clear and transparent mulch:** The transparent film will allow sunlight to pass through and the weeds will grow. The transparent film is quite successful as soil solarization film for disinfecting the soil in order to reduce soil borne diseases and some weeds. This application is quite successful in nursery raising by solarizing the beds before sowing seeds for nursery raising, which gives near 100% seed germination and disease-free nursery. While the black film has proved to be effective in plains to keep crop cool during summer, the transparent film is effective in hilly areas for raising soil temperature in cold climatic conditions during winter (Lyengar *et al.* 2011).

c. **Two side colour mulches:** Wavelength selective or photo-selective films (also called two-side coloured) are designed to absorb specific wavelengths of the sun's radiation, which changes the spectrum of the sunlight passing through the film or being reflected back into the plant canopy. These light changes can have a marked effect on plant growth and development. These films enable growers to control different plant properties such as leaf and fruit size, colour development, yield, plant height, growth, internode length, number of flowering, bloom size, strengthen plant stems, encourage fruit to lower down on plants, and aid in disease control, keeping insects away. The effects are warming of soil temperature, blocking weed growth, increasing colour saturation of developing fruit and increasing carbohydrate transport to developing fruit (Lyengar *et al.* 2011).

Compared to black mulches, wavelength selective mulches reemit less heat, thus maintaining lower leaf temperatures, alter red-far-red light balance leading to phytochrome mediated changes in the plant morphology, and reflect more ultraviolet rays, which repels insects and pests such as aphids, thrips and whiteflies, who transmit viruses. The white/black, silver/black, and aluminized black mulches generally maintain cooler root-zone temperatures, thus suitable for most Indian regions. Effects of some of the coloured mulches are given below:

Available colours:

i. Yellow/black: Attracts certain insects and thus acts as a trap for them, which prevents disease.

ii. White/black: Cools the soil.

iii. Silver/black: Cools the soil, though not to the extent of white/black film and repels some aphids and thrips.

iv. Red/black: Partially translucent allowing radiation to pass through and warm soil, but also reflects radiation back into plant canopy changing ratio of R:FR light, which results in changes in plant vegetative, flower development and metabolism to early fruiting and increased yields in some fruit and vegetable crops.

d. Degradable mulches

i. **Photo-degradable mulch:** This type of plastic mulch film gets disintegrated under sunlight over the designated mulching period.

ii. **Bio-degradable mulches:** This type of plastic mulch film gets disintegrated under natural environmental conditions and gets mixed in soil after mulching period.

2. **Landscape fabric:** Landscape fabric is a better choice for long-term use to suppress weeds because it allows air and water to pass through. It can be used in conjunction with organic mulches and will decompose more quickly than most other inorganic mulches.
3. **Stone:** Examples include volcanic rock, crushed gravel and marble chips. Stones do not retain moisture and can cause heat stress on plants through reflection and ground heating which can burn roots. They are best used away from trees, shrubs, and other plants.
4. **Rubber:** Rubber mulch is composed of recycled or ground tires. This product continues to be researched; however, initial studies indicate possible toxicity levels as well as the risk of flammability. In addition, rubber mulch can remain in the soil indefinitely. It is not recommended for use in the home landscape.
5. **UV-reflective mulches:** These mulches consist of a polythene base to which a thin coat of aluminium ions is adhered. The mulches are collectively referred to as metalized mulches. These mulches reflect UV wavelength which confuses/repels incoming aphid, white flies and leaf hoppers. Therefore, they reduce the incidence of these insects and disease borne by insects like aphid.

6. **White or aluminum reflective mulch:** White or aluminum reflective mulch is used where soil cooling is desired, such as establishing fall crops during the heat of summer. Research has shown that white or aluminum reflective mulch also repels aphids which spread some virus diseases in vine crops such as squash.

Table 4. 3: Thickness of plastic film

Crop recommended	Thickness (in micron)
Annual- short duration crop	20-25
Biennial- mid duration crop	40-50
Perennial-long duration crop	50-100

Source: Lyengar *et al.,* 2011

Table 4. 4: Response of plastic mulch on the yield of vegetable crops

Sl. No.	Crop	Thickness of mulch (micron)	Increase in yield (%)
1	Chilli	25	50-60
2	Potato	25	35-40
3	Cauliflower	25	40-50
4	Tomato	25	45-50
5	Capsicum	25	35-45
6	Okra	25	50-60
7	Brinjal	25	30-35

Source: Lyengar *et al.*, 2011

4.9 Importance of Parameters of the Plastic Film

i. **Thickness:** The early mulch films used were of 60-75 microns (240-300 gauge) thickness, and today it is possible to have 15 micron thick film.

ii. **Width:** This depends upon the inter row-spacing. Normally 1-1.5 m width film can be easily adopted to different conditions.

iii. **Mulch colour:** The colour of the mulch affects soil temperature, temperature of air around the plants and soil salinity.

a. **Transparent film** - Deposits more salt on soil surface.

b. **Black film** - Restricts water movement and upward movement of salt is reduced.

c. **Insect control** - Opaque white film acts as golden colour and attracts insects.

4.10 Selection of Mulch

Rainy season - Perforated mulch.

Orchard and plantation - Thicker mulch.

Soil solarisation - Thin transparent film.

Weed control through solarisation - Transparent film.

Weed control in cropped land - Black film.

Sandy soil - Black film.

Saline water use - Black film.

Summer cropped land - White film.

Insect repellant - Silver colour film

Early germination - Thinner film

Source: TNAU, agriportal

4.11 Precautions for Mulch Laying

i. Do not stretch the film very tightly.

ii. The slackness for black film should be more as the expansion, shrinkage phenomenon is maximum in this colour.

iii. The film should not be laid on the hottest time of the day, when the film will be in expanded condition.

4.12 Methods of Mulching

Mulching area should preferably be equivalent to the canopy of the plant. Required size of mulch film is cut from the main roll. Then clean the required area by removing the stones, pebbles, weeds etc. Till the soil well and apply a little quantity of water before mulching. Small trench could be made around the periphery of the mulching area to facilitate anchoring of the mulch film. Cover the film to the entire area around the tree and the end should be buried in the ground. Semicircular holes could be made at four corners of the film in order to facilitate water movement. The position of the slit/opening should be parallel to the wind direction. Cover the corners of the film with 4-6 inches of soil on all sides to keep the film in position. In hard soil, make a trench of 1' × 1' × 2' depth on four corners of the mulching area and fill it up with gravel or stones, cover the trenches with the mulch film and allow the water to pass through the mulch to the trenches via semicircular holes on the film. Mulch should be laid on a non-windy condition. (TNAU, agriportal)

4.13 Removal and Disposal of Mulch

Non-degradable mulch films should be disposed of properly so as to avoid any plastic pollution problems. In case the mulch film needs to be used for more than one season (thicker film) the plant is cut at its base near the film and the film is removed and used.

4.14 Irrigation

Best way of irrigating mulched crop is drip irrigation. In flood irrigation, channels are made along the un-mulched area and irrigation is carried out. Sprinkler irrigation is also possible.

4.15 Recommendations of Mulch for Specific Fruit Crops

Mango: There was increase in flowering and yield (38–70 % and 40–60 %) of mango cv. Chausa during 'off' and 'on' year by application black plastic mulch (100μ) at CISH, Lucknow. Mulching was done during flower bud differentiation stage (October–November). Paddy straw and dry leaves mulches can be applied uniformly to a thickness of 8 to 10 cm.

Banana: Use of wheat straw and banana leaf sheath as a mulch material (12.5 kg/plant) in banana orchards is useful in increasing the bunch weight and conservation of soil moisture. The mulch is applied at the beginning of summer (February). Black plastic mulch is also very useful. The silver and black colour polythene sheets of 60-micron thickness can be spread over the ridge. Then the edges of polythene should be covered with the help of loose soil.

Pineapple: Black polyethylene mulching with 50-micron thickness is laid down before pineapple plantation to reduce the expenditure on weeding and to conserve the soil moisture and check the soil erosion. Pineapple cultivation with 50-micron black polythene sheet mulching with double row spacing, i.e., 90 × 60 × 30 cm not only suppresses the weed growth but also conserves the soil moisture, which leads to early flowering in mulch plot by 6-7 months.

(https://morungexpress.com/black-polyethylene-mulch-boon-pineapple-cultivation-dimapur).

Citrus: Film mulching significantly increases the quality of citrus fruit, but has negative effect on fruit appearance (Zhang and Xie 2014). They also reported that mulching during the cell division and early-mature stages increased total sugar and reduced sugar content of fruit Statuma mandarin.

Guava: The guava plantation mulched with black polythene produced maximum number of fruits as well as highest yield per plant. Paddy straw mulch was found to be effective to improve the fruit quality of guava. Organic mulches, such as straw, dried grass or compost, are excellent for use under guava trees to eliminate weeds and to conserve moisture (Das *et al.,* 2010).

Aonla: Application of black polythene mulch in aonla during the month of April–June is very effective to conserve moisture. During hot summer months, the crop should be mulched with paddy straw or wheat straw at the base of the tree up to 15-20 cm from the trunk. It conserves moisture as well as controls weed growth.

Litchi: Das *et al.* (2016) observed paddy straw mulch most effective in improving the fruit physico-chemical quality of fruits, leaf mineral composition and biological attributes of soil. The mulches were applied in the month of March around the trunk of the tree to a thickness of 8-10 cm. Mulches were kept in the field for one year.

***Apples, Pear, Plums and Cherries*:** Tree fruit should be mulched to the drip line with 6 to 8 inches of straw or hay in May. Mulch should be reapplied periodically and pulled away from trees at the end of August or early September. This will allow trees to harden off and to reduce mouse pressure around the tree.

***Blueberries*:** Mulch yearly with either wood chips or partially composted sawdust. Plants should be mulched yearly to maintain a 4-foot-wide strip at least 6 inches deep.

***Peach*:** Mulching significantly improve the physical properties of the top soil in peach, which in turn enhanced the hydraulic properties of the matrix, improved crop response, crop growth and fruit yield. Organic mulching techniques in peach cultivation should be considered as a beneficial practice to apply in fruit-trees production under limiting soil conditions (Lordan *et al.* 2015).

***Strawberry*:** Higher yields have been reported for crops such as strawberries (*Fragaria sp.*) when using clear plastic in combination with soil fumigation with methyl bromide and chloropicrin (Johnson and Fennimore 2005). Ali and Gaur (2013) recorded the maximum number of runners per plant, runner plantlets per runner and runner plantlets per plant in strawberry were with paddy straw mulch. Singh *et al.* (2005) observed that the black polyethylene mulch in strawberry had the best growth, fruit weight, yield and quality. Mulching in strawberry protect the flower bud from temperature below 15°F (Tyagi *et al,* 2015). Wang *et al.* (1998) observed that different mulch types had a significant effect on the concentration of ellagic acid in strawberry fruit.

Kiwi: Application of black plastic mulch along with irrigation at 60% of field capacity in kiwi was the best treatment as its effect is nearly similar to that of standard irrigation (Pratima *et al.* 2016).

Nectarine: Andreotti *et al.* (2009) reported that the fruit production and quality of nectarine was enhanced by mulch-treatment.

Olive: Black plastic sheeting and weed-barrier meshes are used under young olive trees. Camposeo and Vivaldi (2011) reported that the olive tree growth in the mulched treatments was equivalent to that in weeded treatments. Yield parameters, both fruit per plant and oil content, also behaved as morphological parameters as well.

References

Andreotti C, Ravaglia D and Costa G (2009). Innovative light management to improve production sustainability, overall quality, and the phenolics composition of nectarine (*Prunus persica* cv. Stark Red Gold). *Journal of Horticultural Science & Biotechnology, ISAFRUIT Special Issue.* 145-149.

Camposeo S and Vivaldi GA (2011). Short-term effects of de-oiled olive pomace mulching application on a young super high-density olive orchard. *Scientia Horticulturae*, **129**: 613-621.

Das BC, Mahi S and Roy MS (2010). Response of soil covers on guava cv. L-49. *Journal of Crop and Weed,* **6** (2): 10-14.

Das K, Sau S, Sarkar T and Dutta P (2016). Effect of organic mulches on yield, physico-chemical qualities and leaf mineral composition of litchi cv. Bombai in Indo-Gangetic plain of West Bengal. *Journal of Crop and Weed,* **12** (3):67-69.

Dong BB, Zhu HT, Zhong ZK and Ye GF (1996). Study on ecological effect of the forest land under-crop sowing and mulching of coastland soil by newly planted. *Acta Agriculturae Zhejiangensis*, **8**: 154-157.

FAO (1990). Protected cultivation in the Mediterranean climate. *FAO Plant Production and Protection Paper.* 90: 313 pp.

Gan J, Zhu Y, Wilen C, Pittenger D and Crowley D (2003). Effect of planting covers on herbicide persistence in landscape soils. *Environ. Sci. Tech,* **37**: 2775-2779.

Ham JM, Kluitenberg G, Lamont W (1993). Optical properties of plastic mulches affect the field temperature regime. *J. Americ. Soc. for Hortl Science,* **118**: 188-188.

Lamont WJ (2005). Plastics: Modifying the microclimate for the production of vegetable crops. *Hort Technology*, **15**: 477-481.

Landis TD (1988). Management of forest nursery soils dominated by calcium salts. *New Forests,* **2**: 173-193.

Lordan J, Pascual M, Villar JM, Fonseca F, Papió J, Montilla V and Rufat J (2015). Use of organic mulch to enhance water-use efficiency and peach production under limiting soil conditions in a three-year-old orchard. *Spanish Journal of Agricultural Research,* **13** (4): e0904, 9 p.

Lyengar SK, Gahrotra A, Mishra A, Kaushal KK and Dutt M (2011). Practical manual on plastic mulching. National Committee on Plasticulture Applications in Horticulture. Department of Agriculture & Cooperation Ministry of Agriculture, Government of India New Delhi, pp: 1- 6.

Mercik S andSmolarz K (1995). Influence of fertilization and mulching on the growth, fruiting and chemical composition of soil and leaves of high bush blueberry. *Acta Hortic.*, **383**: 323-329.

Moniruzzaman M, Mozumde SN and Islam MR (2007). Effect of mulching and pruning on yield and quality of pear. *Bangladesh J. of Agril. Res.*, **32** (2): 225-233.

Pratima P, Sharma N and Sharma DP (2016). Canopy temperature and water relations of kiwifruit cultivar Allison in response to deficit irrigation and in situ moisture conservation. *Current Science*, **111** (2): 375-379.

Schales F (1990). Agricultural plastics use in the United States. *Proceedings of the 11th International Congress of Plastics in Agriculture*, 54-56.

Singh R, Sharma RR and Jain RK (2005). Planting time and mulching influenced vegetative and reproductive traits in strawberry in India. *Fruits,* **60**: 395-403.

Trinka DL and Pritts MP (1992). Micropropagated raspberry plant establishment responds to weed control practice, row cover use, and fertilizer placement. *J. Amer. Soc. Hort. Sci.*, **117** (6): 874-880.

Tyagi S, Singh A, Sahay S and Kumar N (2018). Mulching for Commercial Fruit Production. Chapter-25. Dimensions of Agricultural Sciences. Kalyani Publisher. Pp 292-303.

Wang SY, Galletta GJ, Camp MJ and Kasperbauer MJ (1998). Mulch types affect fruit quality and composition of two strawberry genotypes. *Hort Science*, **33**: 636-640.

Zribi W, Aragüés R, Medina E, and Faci JM (2015). Efficiency of inorganic and organic mulching materials for soil evaporation control. *Soil Till Res,* **148**: 40-45. http://dx.doi.org/10.1016/j.still.2014.12.003.

e-resource

https://morungexpress.com/black-polyethylene-mulch-boon-pineapple-cultivation-dimapur

TNAU, agritech.tnau.ac.in

Outcomes Assessment

Part A

Answer the following questions true or false.

1. Mulching area should be equivalent to the canopy of the plant. True/False
2. The colour of the mulch does not affect soil temperature. True/False
3. In rainy season perforated mulch should be used. True/False
4. Rubber mulch is composed of recycled or ground tires. True/False
5. The black plastic film arrests weed growth completely. True/False

Part B

Answer the following questions.

1. What are organic mulches?
2. Define mulching.
3. What is hydro-mulching?
4. What are different types of plastic mulches?
5. Differentiate organic and inorganic mulches.

Part C

Write a brief note on each of the following.

1. Methods of mulching.
2. Types of inorganic mulch.
3. Selection of mulch.
4. Recommendations of mulches for mango and banana.
5. Advantages of mulching.

5

Horticultural Systems

5.1 Introduction

A system is a group of interacting components, operating together for a common purpose, capable of reacting as a whole to external stimuli; it is unaffected directly by its own outputs and has a specified boundary based on the inclusion of all significant feedbacks. For example, the human body is a system — it has a boundary (e.g., the skin) enclosing a number of components (heart, lungs) that interact (the heart pumps blood to the lungs) for a common purpose (to maintain and operate the living body) (Rana and Rana 2011).The main aim of orchard management in tropical and sub-tropical fruits, is to increase land use efficiency by use of inter-crops, multiple crops, multi-storeyed crops systems etc. in obtaining maximum income from the available land. Apart from these, cover crops, sod culture, herbicides, minimum tillage etc. are also common.

5.1.1 Cropping Systems

It is an important component of a farming system that represents a cropping pattern used on a farm and their interaction with farm resources, other farm enterprises and available technology, which determine their makeup.

Cropping system =Cropping pattern + Management

It is defined as the order in which the crops are cultivated on a piece of land over a fixed period or cropping system is the way in which different crops are grown. In the cropping systems, sometimes a number of crops are grown together or they are grown separately at short intervals in the same field (Rana and Rana 2011). Cropping system in horticulture ensures efficient utilization of available space, increases productivity and profitability. This is particularly more significant for tree fruit crops, which are planted at wider spacing in the field which took period for coming into bearing stage. During these initial periods of orchard establishment, there is wastage of resources like land, water, labour etc. Hence, suitable cropping system should be adopted for maximising yield, net profit and for better utilization of available land, water and labour resources.

5.1.2 Cropping Pattern

It is the pattern of crops for a given piece of land or cropping pattern means the proportion of area under various crops at a point of time in a unit area or it indicates the yearly sequence and spatial arrangements of crops and follows in an area. Fruit based diversified cropping system is now considered to be the most ideal strategy to provide food, nutrition and income security to the people particularly in view of continuous increasing population and decreasing per capita land availability. In addition to contracting land resources, national problems like environmental degradation, climate change, thermal stress, water scarcity, ground water contamination and entry of toxic substances into the food chain are some of the other challenges for the fruit industry. Under such situation, integration of annual crops with fruit trees seems to be a viable option as it would yield multiple outputs, ensure production and income generation in a sustainable manner. Natural resources (land, solar radiation, water, soil) as well as socio-economic inputs (labour, credit, power and market infrastructure) could be efficiently utilised increasing the resilience of the system over time. Integrating annual/seasonal crops with perennial fruit trees/plantation crops approach has been widely recognized by native people, obviously to achieve sustainable production ensuring ecological stability by growing several species together. In the recent past, several innovations and new initiatives have been made both in research and development for improving production and productivity through diversification (Awasthi, 2018).

5.2 Tropical and Sub-Tropical Horticultural Systems

In tropical and sub-tropical fruits, the main aim of orchard management is to increase land use efficiency. Therefore, use of inter-crops, multiple crops, and multi-storeyed crops systems are followed for obtaining maximum income from the available land.

5.2.1 Classifications of Cropping System

Depending on the resources and technology available, different types of cropping systems are adopted on farms, which are (i) Mono-cropping, or Single Cropping and (ii) Multispecies Cropping or Multi-cropping.

5.2.1.1 Mono-Cropping or Single Cropping

Mono-cropping refers to practice of growing only one crop in a piece of land, year after year, e.g., banana or pineapple or any vegetable crops like tomato etc. This is due to climatologically and socio-economic conditions or due to specialization of a farmer in growing a particular crop. The continuous growing of the same species on a piece of land over a sequence of growing seasons leads to risk of residual transfer of pests, diseases, and weeds from one season to the next. This system is common in modern horticulture, where high density planting system is followed. Mostly dwarf or semi-dwarf plants with appropriate canopy management practices lead to ensure better light interception and distribution and ease of mechanization.

However, this system may not be suitable for small growers with marginal holding, who cannot afford to meet high initial cost of inputs and technologies involved and also there is no efficient utilization of land and soil nutrients. There may also be the problem of unfruitfulness due to planting of single varieties of fruit species. Such systems are also criticised for their negative environmental impacts, such as soil erosion and degradation, chemical contamination, loss of biodiversity, and fossil fuel use (Tilman *et al.*2002). Example -planting of a single crop year after year in the same field. In case of monocropping, the field is used to grow only one crop season after season. Here, it is difficult to maintain cover on the soil and it encourages incidence of pests, diseases and weeds and also it can reduce the soil fertility and damage the soil structure.

5.2.1.2 Multispecies Cropping

In case of multispecies cropping, there is large number of compatible crops grown in a same unit of land. It utilises maximum resources efficiently and also ensures better provision of food, fuel, fodder etc and is ideally suitable for small land holdings with maximum production potentiality per unit area and time. In this system, farmers grow two or more crops on farm land in one calendar year with intensive input management practices.

There are various types of multispecies cropping system followed which are described below: (a) Inter-cropping, (b) Mixed cropping, (c) Sequential cropping, (d) Relay cropping and (e) Multi-storey cropping.

A. Intercropping

The term cropping system is often used interchangeably with multiple cropping, which in fundamental nature represents an idea of maximum production per unit area of land within a year or some other relevant time unit with minimum land degradation (Singh, 1972).Therefore, to maximize land use and production, intercropping is an advanced agronomic technique that allows two or more crops to yield from the same area of land. Better utilization of resources and reduced weed competition minimize the risk of food shortages by enhancing yield stability. In this system, farmers grow two or more crops simultaneously on the same field in one calendar year.

Criteria for the selection of an intercrop

1. It should not be tall growing and spreading type.
2. It should not be exhaustive.
3. It should not function as an alternate host for common pests and diseases.
4. Water requirement schedule should match or phenology of crop should match so that operation could be synchronized.
5. It should match for climatic requirements with main plantings.
6. Separate provision for nutrients should be made for intercrop to avoid competition.
7. Intercrops should be preferably legumes or shallow rooted vegetables.

Table 5.1: Some fruit crop-based system with their intercrops

Sl. No.	Crop	Main crop with intercrops	References
1.	Aonla	Aonla + Fenugreek + chick pea + mustard + cumin (Rabi season); Moth bean (Kharif season)	Awasthi *et al.* (2009)
		Aonla + Ber + Karonda + Cluster bean + Brinjal; Aonla + Ber + Karonda + Mothbean + Mustard	Arya *et al.* (2010).
2.	Ber	Ber (Jharber and Bordi) + Mung, Clusterbean, Pearl millet, Mothbean, gram, Vegetable crops	Faroda (1998)
		Ber (*Zizyphus mauritiana*) + Green gram	Gupta *et al.* (2000)
		Ber (Cv. Gola) + Cluster bean+ Mustard + Indian aloe; Wheat + Groundnut	Saroj *et al.* (2003)
		Ber + Wheat; Ber + Chick pea; Ber + Kulthi	Pareek and Nath (1996); Kumar *et al.* (2005)
3.	Mango	Mango + *Phaseolus acutifolius* Genotype "Frijol Escumite" + *Cajanus cajan* (Pigeon Pea) + *Phaseolus vulgaris* Genotype "Frijol de Vara"	Agreda *et al.* (2006)
		Mango + Guava + Cowpea	Nath *et al.* (2007)
4.	Mandarin	Mandarin + Ginger	Upadhyay and Subba (1994)
		Mandarin + Pea	Ghosh *et al.* (1985)
5.	Lemon	lemon + French bean	Vanlalhruaia *et al.* (2013)
6	Banana	Banana +Sweet potatoes	Bekunda and Woomer (1996)

Sl. No.	Crop	Main crop with intercrops	References
7.	Guava	Cauliflower, peach, French bean, cowpea, cluster bean, black gram, okra, onion, turmeric, garlic, cabbage, chillies, papaya	*http:// ecoursesonline. iasri.res.in*
8.	Litchi	Turmeric, ginger, pointed gourd, sweet potato, tomato, radish, cabbage, turnip, brinjal, cucurbits, green gram, black gram, cowpea	
9.	Papaya	Cabbage, cauliflower, chillies, radish, tomato	
10.	Pomegranate	Berseem, Lucerne, cowpea, green pea, cucurbit, cabbage, cauliflower, bean, peas, tomato, carrot, onion, potato, brinjal	
11.	Sapota	Banana, papaya, pineapple, broad bean, tomato, brinjal, cabbage, cauliflower	
12.	Date Palm	*Citrus medica*, guava, Sapota	

Advantages of intercropping

The advantages of intercropping are risk minimization, effective use of available resources, efficient use of labour, increased crop productivity, erosion control and food security (Owuor *et al*, 2002). There is reduction of insect pest populations due to the diversity of crops grown and reduction of plant diseases because the distance between plants of the same species is increased due to the planting of other crops between them, alteration of more beneficial insects especially when flowering crops are included in the cropping system, increase of total farm production and profitability and reduction of weed population through allelopathy and efficient crop production (Maguguda *et al*, 1979). Further, when the intercrop provides a good soil cover, soil temperature will stay relatively low. This prevents burning of the organic matter in the soil and loss of nutrients. It also provides a microclimate that can be favourable for associated crops. Intercropping has important advantages in regard to efficient land use, increasing crop productivity and monetary returns as compared to sole cropping.

Disadvantages of intercropping

1. Decreases in yield as the crops differ in their competitive abilities.
2. Management of intercrops having different cultural practices seems to be difficult task.
3. Higher amount of fertilizer or irrigation water cannot be utilized properly as the component crops vary in their response of these resources.
4. Harvesting is difficult.

Types of intercropping

1. Parallel cropping: Cultivation of such crops which have different natural habit and zero competition. (Black gram + maize)
2. Companion cropping: Such inter-cropping where the production of both intercrops is equal to that of its solid planting. (Potato/Onion + Sugarcane)
3. Multiple cropping: It is one of the cropping systems, wherein two or more crops are grown in succession within a year.
4. Synergistic cropping: The yields of both crops are higher than of their pure crops on unit area basis. (Sugarcane + Potato)

Examples of perennial crop based intercropping system

Intercropping within perennial horticultural crops

Fruit tree and other perennial horticultural crop-based production system can offer suitable option for profitable utilization of the uplands. In case of fruit orchard, intercrops are generally planted in between rows of the main or base crop, with a view to obtain some extra yield without sacrificing the main crop or base crop yield. Fruit based cropping system is a very good option for minimization of crop failure risk at a farm due to anomaly of weather. The intercrops not only generate an extra income but the practice also helps to check the soil erosion through ground coverage and improves the physicochemical properties of the soil.

Mango based intercropping system

Fruit crops like mango and guava constitute a major share of area expansion programme under fruit crops in India. Long juvenile period, heavy mortality of the plants during the summer season due to grazing and lack of irrigation are major factors which discourage the farmers to take up mango orcharding. Again, low productivity of mango under the uplands makes mango cultivation, unprofitable. Hence, development of a profitable mango-based production system with income from the first year onwards can help in alluring the farmers to take up mango orcharding. Intercropping in mango with suitable crops brings good income and improves the fertility of the soil. During the first few years, intercropping can be practiced with no shortage of irrigation. Economic advantages of mango based intercropping systems have been reported by number of workers (Ratha and Swain 2006).

Banana based intercropping system

Bananas are perennial herbs and may be grown on the same piece of land for up to 50 years. Mahant (2011) opined that intercropping in banana was more productive and profitable than their sole cultivation without loss in yield. He reported that intercropping either with onion or garlic in banana at initial growth stage of planting increase the total profit without affecting the yield. Intercropping with garlic and onion with 60% coverage in banana under drip irrigation gave maximum gross and net returns as well as benefit cost ratio.

Citrus based intercropping system

Intercropping of *Citrus* with ground nut, cotton or soybean in kharif season and with wheat or gram in rabi season can be successful. Effective utilization of interspaces between the Sweet orange cv. mosambi plants in the orchard would be a viable and profitable intercropping system for sustainable profit. Ghosh and Pal (2010) obtained highest net return from Mosambi + groundnut combination (Rs. 35,820.0/ha) followed by Mosambi + okra (Rs. 22,520.0/ha) and Mosambi + cowpea (Rs. 22,420.0/ha).

Ber based intercropping system

Ber is planted with wider spacing (5-6 m apart) which provides ideal conditions for growing of field crops in the inter-space. Ber needs pruning every year which is usually done in April-May and it takes about four to five months to develop full canopy which provides unrestricted inter-space during this period. Among the intercropping systems the highest net income (Rs.16,330/ha) and benefit cost ratio (3.85) was obtained in ber + Bengal gram followed by ber + safflower (Yaragattikar and Itnal, 2003). In-situ budded ber cultivar Gola as over storey component and groundnut wheat, cluster bean-mustard and Indian aloe as ground storey component were integrated into the system. The highest B:C ratio (3.67) was obtained in ber + cluster bean-mustard system (Dhaker*et al*. 2013).

Cashewnut-based intercropping

Pattanaik *et al.* (2007) studied the economics of intercrops in cashew plantation with vegetable crops and found that the yield of main crop varied from 60.1 to 8.70 q per ha under various crop combinations. Highest returns of Rs.19,350/- per ha was obtained from colocasia and lowest recorded in cowpea. Net highest return of Rs.44,908/- per ha was obtained when cashew was intercropped with colocasia, followed by brinjal. Minimum net return of Rs.21,350/- per ha was obtained from the sole crop, cashew cv. Vengurla-4.

Coconut based intercropping system

Effective utilization of available space, both horizontally and vertically, is the modern concept of cropping system. Growing coconut as monocrop is not the most efficient way of using natural resources. Adoption of coconut based intercropping system is one of the ways to utilize the natural resources effectively. Adoption of coconut based multiple cropping system emerges as the viable way for improving the economic status of coconut farmers. Intercropping system under coconut is more profitable than monocropping which promises to the farmers a lot besides generating additional employment opportunity (Nath, 2002). Black pepper performed well under coconut (Ghosh, 2009). Tuber crops can be taken profitably in the interspaces of coconut plantation.

Arecanut based intercropping system

Acharya and Singh (2005) in a study on areca nut-based cropping system at CPCRI, Kahikuchi, Guwahati, claimed that the average cost of production of vegetables ranged between Rs. 6750/- and Rs. 12600 /- per ha in case of radish and potato, respectively, whereas the highest gross return of Rs. 42750/- was obtained by growing cabbage followed by brinjal (Rs. 40,600/-) and cauliflower (Rs. 38,000/-).

B. Mixed Cropping

Mixed cropping refers to growing of two or more crops simultaneously with no distinct row arrangement. This cropping pattern is used in vegetable crops, wherein, two or more crops are grown simultaneously without a definite row pattern. Another example of mixed cropping is tomatoes + onions + marigold. It is generally practiced with a view to avoid risk in farming. Here, the seeds of different crops are mixed together and then sown either in lines or they are broadcasted. This system is not scientific and it causes problem in performing all agricultural operations and harvesting of crops. This system of cropping is generally practiced in areas where climatic hazards such as flood, drought, frost etc. are frequent and common. It is a type of subsistence farming and ensures different needs of family. In mixed cropping or diverse cropping two or more crops are grown all at the same time in a field. If by chance one crop fails, the other crops cover the risk of total crop failure. Usually a long duration crop is grown with a short duration one so that both get sufficient nutrition at the time of maturity.

Mixed cropping fulfils diverse functions like complementary use of growth factors, like soil nutrients, light, and water; reduced pest and disease incidence, reduced soil erosion, total biomass production, yield stability and food security.

The intensive agricultural systems involving monoculture is associated with high input of chemical fertilisers and pesticide and also has negative impacts on soil and water quality and on biodiversity conservation. Alternatively, mixed cropping may reveal many potential advantages under various conditions like higher overall productivity, better control of pests and diseases, enhanced ecological balance and greater economic profitability. Numerous studies have shown a significant reduction in harmful insects in mixed cropping systems compared with monocultures of the same species (Vandermeer 1989). A particular type of mixed crop is called trap cropping. Trap crops are those crops which are deployed to attract, divert, intercept targeted insects in order to reduce damage to the main crop (Shelton and Badenes-Perez 2006). Due to mixed cropping, the fertilizer and water requirement of plants are less and pests can be controlled naturally because their natural predators find multiple habitats to survive. It has been found that this practice produces much higher yield per hectare compared to monoculture.

For example, 1) Growing of perennials along with the main crop Arecanut is called Arecanut based mixed cropping system, e.g., Arecanut based mixed cropping systems with banana, cocoa, pepper, betel vine etc.

2) Growing of perennial crops in association with coconut palm is referred to as coconut based mixed cropping, e.g., Cocoa, clove, nutmeg, cinnamon, pepper, betel vine, coffee, and mango.

C. Sequential Cropping

Sequential cropping is a form of multiple cropping in which crops are grown in sequence on the same field, with the succeeding crop planted after the preceding crop is harvested. It refers to the growing of two or more crops in sequence on the same piece of land in the same year. The succeeding crop is planted only after harvest of the preceding crop. Sequential cropping involves growing two crops in the same field, one after the other in the same year. There is no intercrop competition. In some places, the rainy season is long enough to grow two crops: either two main crops or one main crop followed by a cover crop. One example is planting of maize during prolonged rain and then planting of beans during the short rains. Cropping sequences is a rotation system approach in crop production that enables the available natural resources to be preserved and more efficiently utilized. It is the growing of the succession of crops in time on one field in particular time (Matthews *et al*. 2006).

Types of Sequential Cropping

1. Double Cropping

It refers to cultivation of two crops in succession on a piece of land in a year. It involves the harvest of two crops from the same field in a given year. Both economic and environmental concerns motivate the recent interest in double cropping (Searchinger *et al.* 2013). Besides avoiding the negative environmental consequences sometimes associated with cropland expansion, it may offer other environmental or economic benefits. For example, in some applications, double cropping can reduce fertilizer requirements (Heggenstaller *et al.* 2008). Also, by reducing the exposure of soil between harvest periods, double cropping can help protect soil from wind and water erosion. However, in other cases, double cropping may require additional inputs, such as pesticides, herbicides, fertilizer, irrigation water, or use of conventional tillage, which can have negative environmental impacts.

Example: i) Rice – potato/ mustard and ii) Sorghum- gram.

2. Triple Cropping

It refers to cultivation of three crops in succession on a piece of land in a year. According to Ajeigbe *et al.* (2001) triple cropping of wheat-cowpea-rice is viable and sustainable in irrigated cropping systems using early maturing, photo-insensitive cowpea cultivars in parts of Northern Nigeria with supplementary irrigation. Including a legume (cowpea) in the system will benefit not only the subsequent cereal crop and soil but also livestock production in the availability of nutritious feed at the peak of the dry season.

Example: 1) Rice- potato-groundnut, 2) Cowpea- mustard- jute.

3. Quadruple Cropping

It refers to cultivation of four crops in succession on a same piece of land in a year. In the case of quadruple cropping, proper fertiliser, selected quality seeds, right cultivars and plant protection chemicals are used to produce four crops annually and harvest lucrative profits. Cultivation of four crops annually will ensure maximum use of cultivable land, protect soil health, environment and ensure food security despite adverse impacts of climate change.

Example:

i) Kharif groundnut- leafy vegetables- wheat- summer green gram.
ii) Tomato-Ridge gourd-Amaranthus green -Baby corn.
iii) Early Potato-Onion-Mung bean-short duration Rice.
iv) Early Potato-Potato-Mung bean-short duration Rice.

4. **Ratoon Cropping**

Ratoon cropping is a type of sequential cropping where stubbles of the original crop are allowed after harvesting to raise another crop. Ratoon cropping is an old system that has been practiced for many years. Although the origin of ratooning is probably not known for any particular crop, it may have begun when man first noticed the regrowth of new shoots following the cutting of certain crops at harvest, thus, producing a new crop without replanting. The ratoon cropping system consists of mowing the crop, the next crop being the result of the first crop regrowth. This cultural practice is currently used in several crops such as Sugarcane (*Saccharum officinarum* L.) (Bokhtiar and Sakurai 2007), Papaya (*Carica papaya* L.) (Elder *et al.* 2002).

Merits of Ratoon Cropping

1. Low cost of production due to saving in land preparation and after care.
2. Replanting cycle is longer.
3. Higher yield per unit area in a given period of time.
5. Ratoon crop uses less irrigation water and fertilizer than main (original) crop because of a shorter growing period.

Demerits of Ratoon Cropping

1. Lower yield of ratoon crop than the first crop.
2. More incidences of insect pests and diseases.
3. Build-up of harmful weeds.
4. Decrease in crop density (number of plants per unit of land).

D. Relay Cropping

Relay cropping refers to growing of two or more crops on the same field with the planting of the second crop after the first one has completed its development. Relay cropping is a method of multiple cropping where one crop is seeded into standing second crop well before harvesting of second crop. Relay cropping may solve a number of conflicts such as inefficient use of available resources, controversies in sowing time, fertilizer application, and soil degradation. Relay cropping is a complex suite of different resource-efficient technologies, which possesses the capability to improve soil quality, to increase net return, to increase land equivalent ratio, and to control the weeds and pest infestation (Tanveer *et al.* 2017).

For example, in the relay cropping in vegetables, in the beginning two or more crops of different duration vegetable crops are cultivated in the same field. When vegetable of shorter duration is harvested, second vegetable gets better space to grow. Similarly, when second vegetable reaches harvest maturity, a third vegetable is grown and in a similar way the farmers may took 2 to 3 crops in same patch of land over a period of 7 to 8 months.

Advantage of Relay Cropping

Relay cropping compensates the yield loss of main crops by providing additional income from relayed crops. Relay cropping is an efficient crop production system that sustains the income of small farmers through increasing crop productivity and net return per unit area (Yildirim and Guvenc 2005). Relay cropping was proposed as a more profitable system for farmers with reduced risk of crop failure or problems linked with double crop system (Nelson *et al.* 2010). Relay cropping compensates reduction in plant population due to late sowing by giving an opportunity to farmers to sow the intended crop in time. Relay cropping showed impressive results not only in preventing the soil from further deterioration or degradation but also in improving soil health. Inclusion of relay crops may reduce the nutrient losses from the soil leading to less negative impact on environment. Relay cropping possesses a great potential to suppress pest attack either by increasing the number of predators or by disturbing the pest population dynamics (Mensah and Sequeira 2004).

E. Multi-Storeyed Cropping System

Multi-storeyed cropping system (also called multi-tier cropping system) involves combination of plants with various morpho-phenological features to maximize the natural resource use efficiency and enhanced total factor productivity. Two or more crops of different heights are grown simultaneously on a piece of land in a certain period. Sun light, nutrient, land, water and space are utilized in the most efficient and economical way in this system. It improves soil health, soil physical property and enhances higher production stability. Horticulture crops particularly fruit and plantation crops have self-sustainable system where solar energy can be harvested at different heights, soil resources are used efficiently and can increase cropping intensities. The system consists of three main components, *viz*., main crop, filler crop and intercrops which occupy three different tiers in space of the production system. To ensure sustainable productivity and high returns from underutilized and stressed lands and to improve the soil characteristics multi-storey cropping system have been found successful in tropical rain forest, semi-arid and dry land conditions (Nimbolkar *et al.* 2016). Multi-storeyed cropping system is a perspective modern approach for sustainable productivity in horticulture crops in general and plantation crops (coconut, arecanut and coffee) in particular. The system is more adaptable in tall growing perennials with compatible semi-perennial and annual crops (Venkatesh 2015). The practice of multi-storeyed cropping system is highly successful in plantation crops especially in coconut, arecanut, coffee and cashew for enhancing sustainable productivity and realizing higher income per unit area. It is found successful for sustainable productivity in fruit crops (mango, ber, aonla, pomegranate) and plantation crops (coconut, arecanut, coffee, cashew) by which natural resources are utilized efficiently to enhance productivity of main crops (15-20%) and high revenue realization per unit area (50-90%) besides maintaining soil health and balancing environment. Multi-storeyed cropping system involves the growing of crops of different height, rooting pattern, and duration simultaneously on the same piece of land. This is most common in coconut-based cropping system in Kerala to meet diversified needs of farming community for fodder, food and fuel, besides increasing net return per unit area, e.g., coconut+ pepper + pineapple + grass. Taller trees, which has greater requirement for solar light, are able to trap more solar light when grown as top storey and those having requirement of lower light are raised as ground storey. In this system, annual and perennial crops are grown side by side in different tiers by exploiting soil and vertical space more efficiently.

The basic principles of multi-storey cropping system (Nimbolkar *et al.* 2016) are:

(i) Opportunities for crop diversification on scientific, ecological and economic principles.

(ii) Maximize system productivity.

(iii) Utilization of resources with higher efficiency.

(iv) Intensive input use.

(v) Sustainability of farm resources and environment on long term perspective.

Table 5.2: Successful model of multi-cropping system in horticulture-based cropping systems

Cropping sequence	Place of Work	Achievement out of the system	Reference
Coconut + Cocoa + Banana + Moringa + Pineapple	AICRP, Aliyarnagar	This cropping system with 75 % NPK + organic recycling with vermicompost recorded highest nut yield of 182 per palm and highest net income (Rs. 3.80 lakhs per ha) and B:C ratio (2.71)	Nimbolkar *et al.* (2016)
Coconut + cocoa + lime + banana + drumstick	AICRP, Arasikere	With all physical and chemical quality of coconut, this system recorded net income of Rs. 2, 94,810 per hectare compared to monocrop (Rs. 68,200/ha).	Roy *et al.* (2001)
In an area of 1 ha, 150 coconut (7.5 × 7.5 m) + black pepper (1.25 m away from coconut base–150 vines) + cocoa (2.5 m between 2 rows of coconut – 525 plants) + pineapple (1-2 m in the rows, two rows of pineapple-4900 plants)	Coastal region of southern state of India	This model recorded higher yield of coconut (20%) and net returns compared to mono cropping of coconut, besides enhancing soil fertility due to recycling of by-products.	Khan and Krishnakumar (2002)
Mango+ cowpea+ Indian mustard	Todapur Research Orchard of Horticulture Division, IARI, New Delhi	This system recorded significantly highest values of system productivity, gross returns, net returns and benefit: cost ratio during both the crop seasons.	Mirjha and Rana (2016)
Mango + *Phaseolus acutifolius* cv, "Frijol Escumite" + *Cajanus cajan* (Pigeon Pea)	Mango orchards in the Soconusco, Chiapas, Mexico	The biomass of *Cajanus cajan* and *Phaseolus acutifolius* (Frijol Escumite) and their incorporation to the ecosystem produced positive effects on the quality and yield of mango. Likewise, the intercropping of mango with *Cajanus cajan* or *Phaseolus vulgaris* influences the insect diversity in these systems.	Agreda *et al.* (2006)

Cropping sequence	Place of Work	Achievement out of the system	Reference
Mango+ mandarin / Egyptian clover + date palm	-	These systems could be used for the higher net return per unit area and also to combat desertification in sandy soil in arid lands regions.	Abouziena*et al.* (2010)
Mango + cowpea (Kharif) + Bengal gram (Rabi)/ Mango + pigeon pea + tomato	Madhya Pradesh	First system gave maximum productivity per hectare whereas second system came up with maximum monetary returns.	Baghel *et al.* (2003)
(Aonla+ ber + cluster bean+ fennel), (Aonla + bael + cluster bean + Coriander), (Aonla + khejri + cluster bean + ajowain)	Arid regions of Rajasthan	These systems were reported as sustainable and remunerative under the arid ecosystem	Hare Krishan *et al.* (2013)
(Aonla + ber + karonda + cluster bean + brinjal), (Aonla + ber + karonda + mothbean (*Vigna acontifolia Jacq.* Marechal) + Indian mustard (*Brassica juncea*)	Arid regions of Rajasthan	Plant height, number of branches/plant, fruits/ plant and yield was found to be superior in these multispecies cropping systems as compared to sole cropping.	Arya *et al.* (2010)

5.3 Temperate Horticultural Systems

Growth of shoots and roots of fruit trees are influenced by many external conditions like temperature, moisture, soil aeration and availability of nutrients etc. All these factors are greatly influenced by different orchard floor management practices such as cover cropping, inter-cropping, mulching, sod culture, weed management practices, irrigation and fertilizer application. In temperate fruit orchards, the main purpose in their management involves sustainability of soil fertility, control of weed density and soil erosion as well as moisture conservation. These can be achieved through appropriate orchard floor management such as clean cultivation and tillage, cover cropping, mowing of weeds, mulching and herbicide use, and a relatively new method—flame burning of weeds.

5.3.1 Weed Management in Temperate Fruit Crops

Weed control is important in horticultural crops as weeds compete for resources such as water, nutrients, space and light, resulting in reduced crop growth. Some weeds produce allelopathic chemicals to suppress the growth of other plant species surrounding them (Verma and Rao 2006). Weed management in orchards prevents unnecessary competition with fruit bearing trees. So, greater yields and fruit quality are attainable when competition with weeds is eliminated. Orchards are generally infested with various types of annual, biennial and perennial weeds which compete with the fruit plants for nutrients, space, moisture, light and adversely affect growth, vigour, fruit set, flower initiation, yield, fruit quality and winter hardiness (Majek*et al.* 1993) and thereby directly reducing the productivity of fruit trees. Weeds also provide shelter to various pathogens by becoming an alternate or collateral host of invaded crops by a number of fungal, bacterial and viral diseases. It has been reported that about 36-42% losses may occur due to inadequate management of weeds in apple (El-Metwally and Hafez 2007).

Orchard floor management is usually accomplished by various methods, *viz.*, manual, mechanical and chemical means, yet the conventional hand weeding is the most common method. Nevertheless, manual weed control is not only laborious but is also highly expensive. Thus, of late manual and mechanical weed control methods are gradually being replaced by other alternatives such as the use of mulches and herbicides as these are easier, cheaper and less time consuming. The ground management systems studies have shown substantially different effects on soil chemical, biological and physical properties (Laurent *et al.* 2008) as well as differential effects on root-zone microbial communities and tree root development (Yao *et al.* 2005). The combination of mulches and herbicides holds promise as a method for long term control of weeds and reduces labour costs, concomitantly. Weed management strategies for field production requires extensive knowledge of weed biology, herbicide application and calibration procedures, herbicide efficacy against target weeds and correct timing of application (Altland *et al.* 2003). The most common reasons for ineffectiveness of herbicides are improper timing, improper rates, and wrong selection of herbicide for the prevalent weed species. A judicious combination of chemical and cultural methods of weed control would not only reduce the expenditure on herbicides but would benefit the crop by providing proper aeration and conservation of moisture.

5.3.2 Weed Control Methods

5.3.2.1 Cover Cropping and Green Manuring

Cover crops improve soil structure, tilth and water holding capacity and reduce the chance of environmental pollution from nitrogen fertilizers. The weed species are suppressed differently by green manure species, when the green manure is incorporated into the top 5 cm of soil or left on the surface. The emergence of weed seeds is significantly inhibited depending on the species and on the amount and depth of green manure incorporation. Cover crops growing in the alley have been found to compete for nitrogen and available water. If managed properly, cover crops can be beneficial for limiting erosion and increasing soil organic matter (Peregrina *et al.* 2010). Increased soil moisture due to cover crop treatments can decrease soil mechanical resistance and this decrease can lead to increased root density due to ease of root elongation (Becel *et al.* 2012). The legumes generally led to increased trunk size and fruit yield in apple (Sanchez *et al.* 2006). Bhat (2015) observed that plant height, plant girth and plant spread were significantly influenced by intercropping with pea followed by red clover and French bean in sweet cherry trees. Alfalfa and clover mulch and a clover cover crop growing as living mulch increased apple leaf nitrogen in Washington apple orchards (Granatstein and Mullinix, 2008). Pritchard *et al.* (1990) observed, cover crops as an orchard floor management option resulted in greater water infiltration rates in the soil compared with a clean cultivation in almond orchard trees in California. Higher residual effect of green manuring might have increased the soil organic matter content and available nutrients, which in turn, favourably affected fruit yield in the subsequent year.

5.3.2.2 Mulching

Continuous use of organic mulches is helpful in improving the soil physico-chemical properties, microbial flora and soil aeration, which ultimately resulted in better yield and quality fruit. Living mulch residue suppressed pigweed (*Amaranthus* spp.), foxtail (*Setaria* spp.) and velvet leaf (*Abutilon theophrasti*) and maximum weed suppression by living mulch residue occurs shortly after cover crop death. Sharma and Bhutani (1989) observed excellent weed control in peach orchard with 10 cm thick grass mulch. The presence of living mulches helps in controlling escape weeds and may prevent or slow down the invasion of new weeds. Broad leaf weeds were more susceptible to mulching treatments than grassy weeds. Mulches provide effective weed control only if they are applied at proper thickness. Faber *et al.* (2003) found that weed population was least in mulched plot. Mulch thickness of 15 cm resulted in greatest reduction in weeds, but even the 2.5 cm thickness had some effect in Avocado plantation. Application of pine dust mulch to different apple cultivars in the tree row stripes increased mean fruit yield by 19.1-26.3% depending on the cultivar (Solomakhin *et al.* 2012). Kaur and Kaundal (2009) observed lower weed density in pre-emergence application of black polythene mulch and post-emergence glyphosate spray. Whereas, Chandel *et al.* (2010) observed dry grass mulch alone and in combination with post-emergence herbicide (paraquat 1.5 l/ha) is the most effective in controlling weed in kiwi fruit. While, Sharma and Khokhar (2006) observed that black polyethylene significantly increased the plant height, growth, yield and quality of strawberry (*Fragaria x ananassa.*) cv. Chandler, compared to control and other mulches.

Layne *et al.* (2002) used metalized, high-density polyethylene reflective film on apple cv. Imperial Gala at 4 weeks before the first commercial harvest date. They found 27% increase in the red surface compared with the control, without affecting fruit size. Mathieu and Aure (2000) recorded improvement in fruit size due to increased photosynthetic activity of the apple tree cv. Gala in addition to one-week advanced ripening when the plastic reflective mulch was applied after flowering. However, when mulch was applied few days before harvesting it improved the colour of apple fruits. Akasaka and Imai (2002) observed the soluble solids content of the peach fruit juice increased by 8-13% and fruit weight increased by 7-12% per year under reflective mulches. Further the rate of pit splitting also decreased by 10-25%, the concentration of sucrose in the fruit juice increased by 23% but, glucose and fructose concentrations were not affected. So, mulching the orchard with reflective sheet during the maturation period increased the fruit quality of peach.

5.3.2.3 Sod Culture with Herbicidal Strips

The preferred orchard floor management system uses herbicides to maintain a 10 to 12 feet wide vegetation free strip in the tree row with a ground cover in the middle between tree rows. The vegetation free strip eliminates in-row weed competition with trees. The sod or ground cover prevents erosion and serves as an alley to facilitate equipment movement through the orchard during wet weather. The ground cover of choice varies with geographical location.

Some growers prefer to establish a cover crop like winter rye annually in the row middles as opposed to using Bermuda grass, which often grows into the herbicide strip. Regardless of the region, middles need to be mowed several times during the summer. Gormley *et al.* (1973) observed leaf magnesium content increased in Golden Delicious and Cox's Orange Pippin apple trees in association with sod cover. Neilson *et al.* (1984) recorded increased leaf calcium in young Golden Delicious apple tree when found associated with orchard sod cover as compared to grass.

5.3.2.4 Herbicide

Pre-emergence herbicides

Pre-emergence herbicides are applied on the soil soon after tillage operation in spring. Pre-emergence herbicides provide effective control of annual grasses and small seeded broad leaf weeds. It controls germinating weed seeds but usually do not give acceptable control of emerged weeds. Rain fall is needed to properly activate most herbicides; however, best control occurs when water (rain or irrigation) is added within a few days of application. Yadav *et al.* (2015) reported that the major weed flora consisted of *Cyperus rotundus, Echinochloa crusgalli, Commelina benghalensis, Phyllanthus niruri* and *Digera arvensis*andcan can effectively be controlled by pendimethalin + imazethapyr as compared to single application of pendimethalin as pre-emergence and imazethapyr as post-emergence that were equally effective as two hand weeding.

Inclusion ofpre-emergence herbicides and mechanical tillage to eliminate weeds also may reduce soil structure, fertility, and orchard productivity compared with "living" and straw-hay mulches (Merwin *et al.,* 1994). When herbicides are used in vineyards, they provided close to perfect season-long weed control, and with pre-emergent herbicides, not even dead mulch remained. As a result, there is an alarming increase in soil erosion (Hartwig and Ammon, 2002). In pre-herbicides plots with negligible surface vegetation during the summer months, herbicide residues can accumulate and soil physical conditions may deteriorate over time, eventually becoming unfavourable for root growth near the surface. Application of pre-emergence herbicides on mulches can reduce herbicide leaching up to 74% compared to the herbicides applied to the bare soil (Knight *et al.* 2001).

Some pre-emergence herbicides used for controlling weed population are Simazine, Atrazine, Di, Paraquat, Alachlor, Butachlor, Fluchloralin, Pendimethalin etc.

Post-emergence herbicides

Post-emergence herbicides have an advantage over pre-emergence herbicides as the post-emergence herbicides can be selected based on weed species that are present. Glyphosate is an effective herbicide against perennial weeds. Repeated applications of this herbicide have been found effective against *Imperata cylindrica* and also another perennial weed. The continuous use of herbicides might lead to the development and evolution of resistant biotypes. Merwin and Brown (2009) observed greater tree growth and higher yields in post-emergence herbicide treatments in apple orchard.

Weed scouting

It is very important to vigilance the orchards regularly to know the types of weeds and their population in the orchard. Scouting allows growers to kill the escaped weeds with a timely herbicide application. It also helps the farmer for early identification of difficult-to-control weeds. Early identification of problematic weeds can prevent them from becoming established in the orchard. If problematic weeds are noticed for the first time in an orchard, they need to be removed before they produce seed. This can be done by hands or with a spot treatment with a non-selective post-emergence herbicide like Glyphosate and Paraquat. Scouting also gives growers an opportunity to recognize poorly controlled weeds, so their weed management programme can be adjusted. Another aspect grower should consider is the potential for infestation of weeds from around the border of the orchard. Weeds in these areas produce seeds that will find their way into the orchard (Sharma, 2012).

5.3.3 Pollination

Temperate fruit crops are mostly cross-pollinated and pollination is done by honey bee. Cross-pollination is necessary in apple, some pears, plum, peaches, cherry, pecan nut, kiwifruits etc. Most of the apple cultivars grown in India are self-incompatible e.g. Delicious group and they need cross-pollination with a suitable pollinizer cultivar to obtain good fruit set. Honey bees and other native insects are the primary pollinators for apples. However, bees are susceptible to insecticides and they need to be protected at the time of blooming. For this, care should be taken for not to use any insecticide during blooming period, otherwise poor fruiting will occur. Bees are highly attracted to flowers in the ground cover. Before applying any insecticides, remove all weeds (dandelion, clover and other ground cover flowers) present in the orchard by mowing or herbicide. Always follow the safety guideline of the particular insecticide before use and apply it in the early evening to minimize the risk to bees.

5.4 Biological Efficiency of Cropping Systems

Efficient cropping systems for a particular farm depend on farm resources, farm enterprises and farm technology because farm is an organized economical unit. The farm resources include land, labour, water, capital and infrastructure. When land is limited, intensive cropping is adapted to fully utilize available water and labour. When sufficient and cheap labour is available, vegetable crops are included in the cropping systems as they require more labour. Capital intensive crops like banana, pineapple, papaya, turmeric etc. find a space in the cropping system when capital is not a constraint. In low rainfall regions (750 mm/annum) mono-cropping is followed and when rainfall is more than 750 mm, intercropping is practiced. With sufficient irrigation water, triple and quadruple cropping is adopted. When other climatic factors are not limiting farm enterprise like dairy, poultry etc. also influence the type of cropping system. When the farm enterprises include dairy, cropping system should contain fodder crops. Component change in cropping system also takes place with the developments of technology (Rana and Rana 2011). Sustainable horticulture requires adoption of such cropping patterns, which ensure efficient use of available space, enhance productivity and profitability. This is particularly more important for tree fruit crops which are planted at wider spacing in the field and take longer times to come into bearing.

The biological efficiency of a crop or cropping system is measured in terms of yield per unit area over a specified period of time. The yield may be dry-matter yield; biological yield in fruit crops is sum of yields of main product, e.g., fruits or economic yield (i.e. wood, medicinal components etc.). For example, in multi-storeyed cropping system (coconut+ pepper + pineapple + grass); maximum utilization of land and other resources are taken into consideration. In this cropping system, biological efficiency is very high. Farmer gets diverse product (fruit, fodder etc.) as well as secondary/ minor products such as coconut leaf, coconut husk, trunk etc. In inter-cropping (papaya with cabbage/ cauliflower/ tomato), farmer will get both fruit and vegetable. Whereas, in mono-culture, farmer will get only one crop. Thus, the biological efficiencies of cropping systems are governed by and dependent on several factors, most decisive being the type of crops, crop combination, and use, nutrient and water management and agro-climate. So, it is desirable to adjudge the production efficiency of cropping systems by their capacity to produce dry matter per unit area per unit item which is basically determined by genetic makeup of the crop and land.

Some Intercropping System in Perennial Crops

Coconut + Elephant Foot Yam

Multi Cropping system in Coconut

Cashew + Turmeric

Aonla + Guava+ Urd bean

Mango+ Pine apple

Mango + Chilli

Mango + Radish

Intercropping in Banana orchard

Multistoreyed cropping system (Khejri+ Ber + Grasses), Source: CAZRI

References

Abouziena HFH, Elham Z, Abd El-Motty, Youssef RA and Sahab AF (2010). Efficacy of intercropping mango, mandarin or Egyptian clover plants with date palm on soil properties, rhizosphere microflora and quality and quantity of date fruits. *J. American Sci.*, **6** (12): 230-238.

Acharya GC and Singh LS (2005). Arecanut based cropping system: alternative pathway to achieve sustainability in North Eastern India. *Indian J. Arecanut, Spices and Medicinal Plants*, **16** (1): 23-26.

Agreda FM, Pohlan J, Marc J and Janssens J (2006). Effects of Legumes Intercropped in Mango Orchards in the Soconusco, Chiapas, Mexico. *In: Conference on International Agric. Research for Development*, 1-6.

Ajeigbe HA, Muhammad HI and Singh BB (2001). *Potential of Triple Cropping of Wheat-Cowpea-Rice with Supplemental Irrigation in Sudan Savanna Zone of Nigeria*. International Crop Research Institute for Semi-Arid Tropics, Kano Station.

Arya R, Awasthi OP, Singh J and Arya CK (2010). Comparison of fruit based multi-species cropping system under arid region of Rajasthan. *Indian J. Agric. Sci.*, **80** (5):423-426.

Altland JE, Wehtje G and Gilliam CH (2003). Weed control in field nurseries. *Hort Technology,* **13**: 9-14.

Awasthi OP, Singh IS and More T (2009). *Indian Journal of Agricultural sciences,* **79** (8): 587-591.

Awasthi OP (2018). Fruit based diversified cropping system: an alternative approach for nutritional and economic security. *Progressive Horticulture*, **50** (1& 2): 41-46.

Baghel BS, Tiwari R and Gupta N (2003). Productivity and profitability of mango based intercropping system under rainfed agro-climatic conditions of Madhya Pradesh. *South Indian Hort.*, **52** (1-6):1-4.

Becel C, Vercambe G and Pagés L (2012). Soil penetration resistance, a suitable soil property to account for variation in root elongation and branching. *Plant and Soil*, **353**: 169-180.

Bekunda M and Woomer PL (1996). *Agric. Ecosys. Environ.*, **59** (3): 171-180.

Bhat R (2015). Response of young "Misri" sweet cherry trees to orchard floor management. *International Journal of Advanced Research*,**3** (5): 638-648.

Bokhtiar SM and Sakurai K (2007). Effects of integrated nutrient management on plant crop and successive first and second ratoon crops of sugarcane in Bangladesh. *J. Plant Nutrition*, **30**: 135-147.

Chandel JS, Chauhan N and Singh R (2010). Effect of orchard floor management systems on growth, yield, fruit quality and nutrient contents of kiwifruit (*Actinidia deliciosa* Chev.). *Horticultural Journal*, **23** (1): 1-5.

Dhakar MK, Sharma BB and Prajapat K (2013). Fruit Crop-Based Cropping System: A Key for Sustainable Production. *Popular Kheti*, **1**(2): 39-44.

El-Metwally IM and Hafez OM (2007). *Arab University Journal of Agricultural Science, Ain Shams University, Cairo* **15** (1): 157-166.

Elder J, Reid DJ, Macleod WNB and Gillespie RL (2002). Post ratoon growth and yield of three hybrid papayas (*Carica papaya* L.) under mulched and bare ground conditions. *Australian J. Exp. Agric.*, **42**: 71-81.

Faber BA, Downer AJ and Menge JA (2003). Effects of mulch on avocado and citrus. *Proceedings V World Avocado Congress (Actas V Congreso Mundial del Aguacate)* pp. 719-724.

Faroda AS (1998). Arid zone research: an overview. (In) Faroda AS and Manjeet Singh Ed. *Fifty Years of arid zone research in India*. CAZRI, Jodhpur. pp. 1-16.

Gahukar RT (1989). Pest and disease incidence in pearl millet under different plant density and intercropping patterns. *Agriculture, Ecosystems & Erosion*, **26** (1) : 69-74.

Ghosh D, Mitra SK and Bose TK (1985). Fruit growing in India. In: *Proceeding of the 3rd National Citrus Seminar*, Calcutta, India, pp. 1-2.

Ghosh DK (2009). Performance of black pepper (*Piper nigrum* L) as an intercrop with coconut in the alluvial plains of West Bengal. *Indian Coc. J.* **51**: 4-7.

Ghosh SN and Pal PP (2010). Effect of intercropping on plant and soil of Mosambi sweet orange orchard under rainfed conditions. *Indian J. Hort.*, **67** (2):185-190.

Gormley R, Robinson D and Okennedy N (1973). The effect of soil management systems on the chemical composition and quality of apples. *Journal of the Science of Food and Agriculture*,**24**: 227-239.

Granatstein D and Mullinix K (2008). Mulching options for northwest organic and conventional orchards. *Hort Science*, **43**: 45-50.

Gupta JP, Joshi DC and Singh BB (2000). Management of arid agrosystem: In: National resource management for agriculture production in India (Eds) JSP Yadava and GB Singh). International conference on managing natural resource for sustainable agriculture production in the 21st Century.

Hance RJ and Holly K (1990). *Weed Control Handbook: Principles*. Blackwell Scientific Publications, Oxford, London.

Hare Krishan, Singh IS, Bhargava R and Sharma SK (2013). Fruit based cropping systems for sustainable production. *ICAR News*, **9** (2): 9

Hartwig NL and Ammon HU (2002). Cover crops and living mulches. *Weed Science*, **50**: 688-699.

Heggenstaller AH, Anex RP, Liebman M, Sundberg DN and Gibson LR (2008). Productivity and Nutrient Dynamics in Bioenergy Double-Cropping Systems. *Agronomy J.*, **100**: 1740-1748.

Kahn BA (2001). A ratoon cropping system for fall bell pepper production. *Hort. Sci.*, **36**: 897–899.

Kaur K and Kaundal GS (2009). Efficacy of herbicides, mulching and sod cover on control of weeds in plum orchards. *Journal of Weed Science,* **41**(12): 110-112.

Khan HH and Krishnakumar V (2002). Spices in coconut based cropping system. In: Proceedings of the National Seminar on Strategies for increasing production and export of spices, Calicut, 1-17.

Knight PR, Gilliam CH, File SL and Reynolds D (2001). Mulches reduce herbicide loss in the landscape. *Proceeding Southern Nursery Association Research Conferences* **46**: 461-463.

Laurent SA, Merwin IA and Thies JE (2008). Long-term orchard groundcover management systems affect soil microbial communities and apple replant disease severity. *Plant and Soil,* **304**: 209-225.

Layne DR, Jiang Z and Rushing JW (2002). The influence of reflective film and Retain on red skin coloration and maturity of "Gala" apples. *Hort Technology,* **12**: 640-645.

Mahant HD (2011). Intercropping studies in banana cv. Grand Naine under drip irrigation. *A thesis submitted to faculty of Agriculture NAU,* Navsari.

Majek BA, Neary PE and Polk DF (1993). Smooth pigweed interference in newly planted peach trees. *Journal of Production Agriculture,* **6**: 244-246.

Mathieu VD and Aure R (2000). Apple, a reflective film for improving colour. *Infos Paris,* **160**: 37-41.

Matthews KB, Buchan K, Sibbald AR and Craw S (2006). Combining deliberative and computer based methods for multi-objective land use planning. *J. Agricultural Syst.*, **87**: 18-37.

Mensah RK and Sequeira RV (2004). Habitat manipulation for insect pest management in cotton cropping systems. *In: Gurr GMS, Wratten D, Altieri MA (eds) Ecological engineering for pest management: advances in habitat manipulation for arthropods.* Cornell University Press, New York, pp. 187–197.

Merwin IA and Brown G (2009). Apple root growth, turnover and distribution under different orchard ground cover management systems. *Hort Science,* **44** (1): 168-175.

Merwin IA and Stiles WC (1994). Orchard ground cover management impacts on apple tree growth and yield and nutrient availability and uptake. *Journal of American Society for Horticultural Science,* **119**: 209-215.

Nath JC (2002). Prospects of coconut based high density multi-storeyed cropping in Assam. *Indian Coc. J.*, **33**: 10-11.

Nath V, Das B, Yadav MS, Kumar S and Sikka AK (2007). Guava - a suitable crop for second floor in multi-storied cropping system in upland plateau of eastern India. *Acta Horticulturae*, **735**: 277-295.
(DOI:10.17660/ActaHortic.2007.735.39,https://doi.org/10.17660/ActaHortic.2007.735.39).

Nelson KA, Meinhardt CG and Smoot RL (2010). Wheat (*Triticum aestivum* L.) cultivar selection affects double-crop and relay-intercrop soybean (*Glycine max* L.) response on clay pan soils. Hindawi Publishing Corporation, *International Journal of Agronomy*, Volume 2010, Article ID- 543261, 8 pages, (DOI:10.1155/2010/543261).

Nimbolkar PK, Awachare C, Chander S, Husain F (2016). Multi Storied Cropping System in Horticulture-A Sustainable Land Use Approach. *Int. J. Agric. Sci.*, **8** (55): 3016-3019.

Owuor C, Tenywa JS, Muwanga S, Woomer PL And Esele P (2002). Performance of a sorghum-legume intercrops in response to row levels of nitrogen. Program and extended abstracts. *FORUM Working Document. No. 5. The Forum on Agriculture Resource Husbandry,* Nairobi, Pp. 542-544.

Pattnaik AK, Lenka PC, Mohapatra KC and Mohaptra RN (2007). Intercropping in the cashew plantation with vegetable crops, *National Seminar on Research, Development and Marketing of Cashew.* 20th – 21st November 2007:61.

Peregrina F, Larrieta C, Ibanez S and Garcia-Escudero E (2010). Labile organic matter, aggregates and stratification ratios in a semiarid vineyard with cover crops. *Soil Science Society of America Journal,* **74**: 2120-2130.

Pritchard TL, Asai WK and Hendricks LC (1990). Orchard vegetation management: Effects on water use and soil characteristics. *Western Pecan Growers Association Proceeding* pp. 56-62.

Rana SS and Rana MC (2011). *Cropping System.* Department of Agronomy, College of Agriculture, CSK Himachal Pradesh Krishi Viswavidyalaya, Palampur, 80 pages.

Ratha S and Swain SC (2006). Performance of intercrops in mango orchard in Eastern Ghat High Land Zone of Orissa. *Indian J. Dry land Agric. Res. Dev.*, **21**(1): 12–5.

Roy S, Raj S, Choudhury M, Dey SK, and Nazeer MA (2001). Intercropping of banana and pineapple in rubber plantations in Tripura. *Indian J. Natural Rubber Res.*,**14** (2):152-158.

Sanchez EE, Giayetto A, Cichon L, Fernandez D, Aruani MC and Curetti M (2006). Cover crops influence soil properties and tree performance in an organic apple (*Malus domestica* Borkh) orchard in northern Patagonia. *Plant and Soil,* **292**: 193-203.

Searchinger T C, Hanson J, Ranganathan B, Lipinski R, Waite R, Winterbottom A and Heimlich R (2013). *Creating a Sustainable Food Future: Interim Findings.* Washington, DC: World Resources Institute. Available at: http://www.wri.org/publication/creating-sustainable-food-future-interim-findings

Sharma CL and Khokhar UU (2006). Effect of different mulches and herbicide on growth, yield and quality of strawberry (*Fragaria x ananassa* Duch.) cv. Chandler. In: *Temperate Horticulture: Current Scenario* (Eds. DK Kishore, SK Sharma and KK Pramanick), New India Publishing Agency, New Delhi pp. 313-320.

Sharma N (2012). *Orchard Soil Management. E-course*. YSPUHF.

Sharma YP and Bhutani VP (1989). Effect of nitrogen levels and weed control treatments on nutrient removal by weeds and leaf nutrient status of peach (*Prunus persica*). *Indian Journal of Agricultural Sciences* **59**: 255-259.

Shelton AM and Badenes-Perez FR (2006). Concepts and applications of trap cropping in pest management, *Annu. Rev. Entomol.*, **51**: 285–308.

Singh A (1972). Conceptual and experimental basis of cropping patterns. *In*: *Proceedings of symposium on Cropping Patterns in India.* ICAR, New Delhi.

Singh SP (1997). *Cropping systems in vegetable crops, principles of vegetable production.* Agrotech Publishing Academy.

Solomakhin AA, Trunov YV, Blanke M and Noga G (2012). Organic mulch in apple tree rows as an alternative to herbicide and to improve fruit quality. *Acta Horticulturae*, **933**: 513-521.

(DOI: 10.17660/ActaHortic.2012.933.6).https://doi.org/10.17660/ActaHortic.2012.933.67)

Tanveer M, Anjum SA and Hussain S (2017). Relay cropping as a sustainable approach: problems and opportunities for sustainable crop production. *Environ Sci Pollut Res.*, **24** (8): 6973–6988.

Tilman D, Cassman K, Matson P, Naylor R and Polasky S (2002). Agricultural sustainability and intensive production practices, *Nature*, **418**: 671–677.

Upadhyaya RC and Subba JR (1994). Annual Report 1994. Department of Horticulture, Government of Sikkim, Krishi Bhawan, Tadong, Sikkim.

Vandermeer J, Van Noordwijk M, Anderson J, Ong C and Perfecto I (1998). Global change and multi-species ecosystems: concepts and issues, *Agr. Ecosyst. Environ*, **67**: 1–22.

Vanlalhruaia H, Gopichand B and Lalnunmawia F (2013). *Sci Vis.*, **13**(1): 40-44.

Venkatesha J (2015). Multi storied cropping system in Horticulture- A Perspective Approach for Sustainable productivity. *In: Proceedings of winter school on multistoried cropping system and canopy architecture management in horticultural crops held* from 28th Sept. to 18th Oct. 2015, College of Horticulture, Sirsi, Karnataka, pp. 127-133.

Verma M and Rao PB (2006). Allelopathic effect of four weed species extracts on germination, growth and protein in different varieties of *Glycine max* (L.) Merrill. *Journal of Environmental Biology,* **27** (3): 571-577.

Yadav KS, Dixit JP and Prajapati BL (2015). Weed management effects on yield and economics of blackgram. *Indian Journal of Weed Science,* **47** (2): 136-138.

Yao S, Merwin IA, Bird GW, Abawi GS and Thies JE (2005). Orchard floor management practices that maintain vegetative or biomass groundcover stimulate soil microbial activity and alter soil microbial community composition. *Plant and Soil,* **71**: 377-389.

Yaragattikar AT and Itnal CJ (2003). Studies on Ber Based Intercropping Systems in the Northern Dry Zone of Karnataka. *Karnataka J. Agric. Sci.,* **16** (1): 22-25.

Yildirim E and Guvenc I (2005). Intercropping based on cauliflower: more productive, profitable and highly sustainable. *European J. Agron.,* **22**: 11–18.

e-resource

http://ecoursesonline.iasri.res.in

Outcomes Assessment

Part A

Answer the following questions true or false.

1. Temperate fruit crops are mostly cross-pollinated. True/False
2. Simazine is a post-emergence herbicide. True/False
3. Cover crops do not improve soil structure, tilth and water holding capacity. True/False
4. Multi-storeyed cropping system is also called multi-tier cropping system. True/False
5. Relay cropping does not compensate the yield loss of main crop. True/False

Part B

Answer the following questions.

1. What is intercropping?
2. Differentiate between intercropping and mixed cropping.
3. Cite two merits of ratoon cropping.
4. Define relay cropping.
5. Differentiate between pre-and post-emergence herbicides.

Part C

Write a brief note on each of the following.

1. Biological efficiency of cropping systems.
2. Basic principles of multi-storey cropping system.
3. Types of sequential cropping.
4. Cropping pattern.
5. Cover cropping and green manuring.

6

Systems of Irrigation

6.1 Introduction

Irrigation is the process of applying water to soil, primarily to meet the water requirement of plants. Irrigation is mostly done when rain water availability in the soil for plant is insufficient. Water from different sources like rivers, lakes, reservoirs or aquifers is pumped or flows by gravity through pipes, canals, ditches or even natural streams. Irrigating the fields enhances the quantity, quality and reliability of crop production. According to the Food and Agriculture Organization of the United Nations, irrigation contributes to about 40% of the world's food production on 20% of the world's cultivated land (Bjorneberg, 2013).

Though nearly 75% of the total area on the earth's surface is occupied by water, 97.3% of the water is in oceans and seas, and only 2.47% is the fresh water. Water is a vital factor and natural resource for the growth of any living organism, and is a limiting factor for plant growth. Plants require water for their better growth and production. Where there is shortage of water, particularly during critical stages like flowering and fruiting, there can be drastic reduction in yield. Hence, the necessity of irrigation is to make up this deficiency of water.

Many horticultural crops are dependent upon stored soil water during rainfall season and on irrigation during the summer period. Therefore, accurate determinations of irrigation methods, timing and volumes are essential if sustainable horticultural development and environmentally sound water management are to be achieved. Sustainable irrigation, which includes the application of the regulated deficit irrigation and specific crop coefficients to calculate the plant water requirement, reduces irrigation induced salinization risk and increases yield and quality. To choose the most appropriate irrigation method and design, one must know the soil characteristics, water requirements of the crop, water availability and water quality.

A sound knowledge of crop characteristics such as soil volume explored by roots, their sensitivity to water stress and their seasonal water requirements, are of primary importance. This information is required not only to improve understanding the underlying processes of plant physiology and their control (Dragoni *et al*., 2005), but also for improving the design of irrigation systems and irrigation scheduling. Due to the scarcity of water resources, accurate evaluation of water use efficiency by the different crops is also very important. However, in spite of the large number of methods to measure or estimate plant water use (Rana and Katerji, 2000) further efforts are required to improve our understanding of crop water use efficiency.

Except in soils of low water-holding capacity, localized irrigation methods (drip irrigation or sub-irrigation) are best for all fruit tree species. However, in the case of kiwifruit (*Actinidia* spp.) because of its physiology and its root system characteristics (Xiloyannis *et al*. 1993) irrigation methods that wet the whole soil surface should be considered instead.

Management of irrigation is an important practice in fruit production regions of the world, particularly in arid and semi-arid climates. It draws attention to the principles of irrigating fruit crops, and identifies the critical factors for measuring crop requirements and the control of irrigation, both to optimize cropping and minimize waste. Frequency of irrigation and amount of water applied to fruit crops is a common concern among the fruit producers of India. Insufficient water supply may result in reduced tree growth, yield and fruit quality due to water stress. Excessive irrigation, on the other hand, may increase nutrient leaching, water-logging problems, incidence of pests and diseases and the associated cost of frequent operation and maintenance of the irrigation system. Optimizing water applications by scheduling irrigation to fruit orchards may increase water productivity, reduce production costs, and increase tree growth, fruit yield and quality.

6.2 Soil Water-Holding Capacity

The soil can store huge amounts of water. In particular, rain water accumulates in autumn and in winter when plant water use rates are low. Deep, loamy soils can hold up to 2,000 m^3 per hectare if a 1 m rooting depth is assumed. The water contained in such a volume of soil is sufficient to meet about 30-40% of an orchard's annual water requirement. In light, shallow soils, and in areas having a shallow water table where the root systems cannot develop to a very great depth, the amount of water that can be stored in the soil is much more limited and, consequently, plants are more likely to be exposed to water-deficit induced injury in the summer period (Xiloyannis *et al.* 2012). Under such water scarcity conditions, soil management should aim at: (i) improving the soil's water holding capacity during rainfall season and (ii) reducing soil surface evaporation and transpiration from fruit trees and cover crops.

Water loss by surface evaporation can be as high as 50% of precipitation and can amount to about 30% of yearly evapo-transpiration. Soil surface evaporation losses increase with decreasing of the Leaf Area Index (LAI) (m^2 of leaf per m^2 of soil), and with increasing numbers of irrigation events, especially if using methods that wet the whole soil surface. The distribution efficiency of the various irrigation methods applied to full bearing orchards varies from 50 to 90%, because of the different amounts of water evaporating from the soil between irrigation events and also during distribution. In young orchards, where the root systems are not fully developed and where ground cover is limited, the efficiency of the various irrigation methods varies from between 10 to 95% (Xiloyannis *et al.*, 2012).

6.3 Soil-Water- Plant Relationships

This relationship plays a vital role in productivity of orchard. Soil holds the water and act as reservoir of water to be used by plants through their roots present in the soil. Water is required by the plants for their metabolic activities, maintaining cell turgidity, as well as an important constituent of cells and more important is a carrier of nutrients. Proper air circulation/ air space must be in soil, otherwise it hampers root respiration which may cause rotting of root. The rate of entry of water and air into the soil and its retention, movement and availability to roots are all physical phenomena. Therefore, before irrigating the land, farmer should realize the importance of this relationship so that he can choose best irrigation system for his crop. A good soil should contain 25 % air, 25 % water and 50 % soil (45% mineral matter and 5% organic matter).

6.4 Systems of Irrigation

Various irrigation methods have been developed over time to meet the water requirement of certain crops for specific areas. Broadly, there are three main methods to apply water to the plants: (1) surface irrigation, (2) sub-surface irrigation and (3) Micro-irrigation. Each system has advantages and disadvantages. Physical factors involved in the system selection include soils, climate, topography, water quality and availability, water table depth, orchard size, system performance, maintenance and repair etc. Human and economic factors involved are labour and management skills, availability and cost, as well as capital and energy costs in relation to expectable returns.

No system is best for all situations. Sprinkler and micro-irrigation are often better choices than surface irrigation on sandy soil where excessive percolation is a problem. Surface irrigation may be better in arid, windy areas where wind and evaporation losses are significant. Surface irrigation offers less control of application depth; so small, frequent irrigations are not practical for water sensitive crops, which are better suited to sprinkler irrigation. According to the International Commission on Irrigation and Drainage, surface irrigation is used on about 85% of the 299 M ha of irrigated crop land in the world. India and China each irrigate more than 60 M ha of crop land, accounting for almost half of the irrigated land in the world (FAO 2013). Approximately 95% of the irrigated land is surface irrigated in India and China (Bjorneberg 2013).

6.4.1 Surface Irrigation

Surface irrigation has five main techniques, i.e., (i) Flooding (ii) Basin irrigation (iii) Border irrigation (iv) Furrow irrigation and (v) Pitcher irrigation. All these methods are employed to irrigate fruit orchards. This type of irrigation is recommendable only in those regions where low-cost labour and abundant water are available or where rainfall distribution is adequate to meet crop requirements and only occasional irrigation is necessary. It is suited to land that is almost flat.

Surface irrigation entails water flowing by gravity over soil. Water is usually supplied by gravity or pumped from the water source through canals, pipes or ditches to the field. In some locations, however, water may need to be pumped from the source to a field at a higher elevation. Types of surface irrigation systems include flooding, basin, check basin, border irrigation and furrow. Surface irrigation systems are typically used for field crops, pastures and orchards. Efficiency of surface irrigation systems vary tremendously because of variations in soil type, field uniformity, crop type and management. Surface irrigation is often considered less efficient than micro-irrigation system. However, a well-managed surface irrigation system on a uniform soil with a runoff reuse system can approach 90% application efficiency (Bjorneberg 2013).

6.4.1.1 Flooding

It is the least controlled of all surface irrigation techniques. Water is delivered to the orchard and allowed to spread over the entire orchard surface in a continuous sheet. Since the degree of control over the flowing water is minimal, the resulting distribution of water is usually highly uneven. Despite its obvious disadvantages, flooding is still prevalent in many areas for irrigation of orchards. In this system, whole orchard area is irrigated through one head, i.e., without sub-division of the unit area into small plots. Follow this method if the land is flat and levelled and plenty of water is available. The method is useful where intercrops or green manuring crops are grown in the orchard. In this system, wastage of water is more; weed growth is excessive, risk of bark diseases like collar rot or foot rot exist as the tree trunk remain in contact with water for a longer time. It is suitable for orchard attaining more than 8-10 years of age.

Advantages

1. The entire soil gets fully saturated with water without leaving any dry pocket in it.
2. As involvement of labour is less, the operational cost is also less in this method.
3. Rodents are controlled to considerable extent.
4. By flooding trees are well protected from frost as the latent heat of water is more.

Disadvantages

1. The orchard trees badly suffer due to over-irrigation if the irrigation water is not drained off quickly and timely. It may create anaerobic condition around.
2. Huge amount of water is wasted. Water use efficiency is low.
2. The method cannot be used unless the land is even.

6.4.1.2 Basin Irrigation

In this system, low earth dikes are made around each tree or a group of trees, within which water can be impounded for irrigation. Young fruit plants are usually irrigated by basin irrigation because at this stage the plants are irrigated individually through a channel connecting the basins made around their trunks. Old trees can also be irrigated by this method in those areas where the natural land surface is nearly level. In sloppy land, basin irrigation can be carried out in conjunction with terracing.

Basin size should be a balance of soil infiltration rate, slope and water supply. Water depth in basins varies from about 5-20 cm, with typical depths of 10-15 cm. Efficient basin irrigation requires a level soil surface with uniform soil texture and adequate water supply so the basin is quickly and uniformly covered with water. If the basin is not level, the higher elevation areas will receive less water than the low areas. If the basin inflow rate is inadequate, water will slowly advance, causing large differences in infiltration opportunity time within the basin (Fangmeier *et al.* 1999).

Advantages

1. Very suitable to loose-textured soil where flooding or furrow method has limitation.
2. Water stagnation in the basin dose not takes place as the excess water flows to the next basin and is finally drained off.

Disadvantages

1. The nutrients present in the soil in a basin are washed away and move to the next basin.
2. The soil-borne pathogens, if exist in the soil of a basin are carried to the other basins and thereby, the soil of those basins is also contaminated.
3. Wastage of water occurs.

Check Basin: In this method, the field is divided into small plots/ blocks surrounded by small bunds on all the four sides. Water from head channel is supplied to the field channels one after another. Each field channel supplies water to two rows of check basins and water is applied to one basin after another. This method is adopted when the field is quite large and is not easy to level the entire field. In such situations, the field is divided into small strips and each strip into several plots/blocks by putting bunds and those plots are called check basin. Each plot is levelled for uniform water application. Attention should be paid to block the sub-channels after basin has received adequate water. This is a good system of irrigation for young orchard up to 6-8 years of age and also for arid regions, having water shortage.

Advantages

1. More uniform in distribution of water.
2. Over-irrigation chances are less as the entry of irrigation water in the beds is regulated.
3. The soil is made to saturate more evenly.

Disadvantages

1. Unless the land is levelled, distribution of water in the plot is uneven.
2. Considerable area is lost under field channels and bunds, i.e., nearly 5 % of area.
3. More labour is required for field layout and irrigation.
4. Bunds interfere in working of inter-cultivation equipment.

***Modified Basin System*:** This is an improved system over the basin system. In this system, main channel runs in between two tree rows and the basins are linked independently through small sub-channels. Attention should be paid to block the sub channels after basin has received adequate water. The size of basin is increased with the increase in tree canopy size every year. This is a good system of irrigation for young orchard up to 6-8 years of age and also for arid regions, having water shortage. Incidence of bark diseases are less and intercropping is possible.

Advantages

1. Weeds can easily be controlled by ploughing, as alternate rows do not have any channels.
2. Nutrients from one plant to other do not move.
3. Any disease or pathogens also do not transmit from one plant to other plant.

Disadvantages

1. More labour is required to prepare and maintain the channels. So, cost of production increases.

6.4.1.3 Border Irrigation

It is followed in large levelled fruit orchards where the slope of the surface does not exceed 0.5%. Here the entire orchard area is divided into elongated plots/long strips confined to low earth bunds. Depending upon the slope and the size of the land, the number of border strips can be decided. This method is also called as the Sara method. In this system, water flows between dikes rather than ponded within basins. The irrigated areas between dikes may be 3-30 m wide and up to 400 m long. The field slope between dikes (perpendicular to water flow direction) should be nearly level so water flows uniformly down the field.

6.4.1.4 Furrow irrigation

In this system, the water runs in small furrows between the tree rows. The furrows vary from 30 to 60 cm in depth, 20 to 60 cm in width and 200 to 300 cm in length and follow the contours of the land. Slopes along the furrow may range from 0.2 to 2% but sometimes may be as high as 5%. This system is more suitable for irrigating young fruit orchards as one furrow running alongside of each tree row may suffice to supply water to the limited zone of rooting. When the trees develop, several furrows are needed between the rows in order to wet the greater volume of soil occupied by the extensive root system. This irrigation is very convenient for irrigating pineapple, banana and papaya orchard.

Water commonly flows in furrows for 12-24 hours during irrigation, but shorter or longer durations may be used depending on furrow length, soil properties, and water management consideration. Inflow rates for individual furrows can vary from about 10 to 100 L per min., again depending on soil, slope, field length and management considerations. Ideally, water should advance across the field in about 25% of the total irrigation time to uniformly irrigate the field. Since soil erosion increases as field slope and inflow rate increase, flow rate must be carefully managed on fields with steeper slopes (>1%). Low inflow rates and long irrigation durations may be needed to apply the desired amount of water during irrigation on soils with low infiltration rate. Conversely, higher inflow rates are often needed on fields with low slopes and/or high infiltration rate soils in order for the water to flow across the field and uniformly irrigate the upper and lower portions of the field (Bjorneberg, 2013).

Furrow irrigation requires lower capital investment, less technical knowledge and greater labour than most other irrigation systems. Fields can be irrigated without levelling or grading because water flows in furrows. Furrow irrigation is not well suited to automation because water flow rate must be adjusted for each furrow for each irrigation (Bjorneberg, 2013).

Advantages

1. Evaporation loss of water is less.
2. The irrigation water is able to reach greater depth of the soil.
3. For such fruit crops which are needed to be earthed up occasionally as in banana, the method is preferable. This is because, earthing-up is done automatically while opening the furrows.
4. Where the soil is of higher clay content, the method is considered to be particularly suitable.

Disadvantages

1. The method is not suitable in porous soil for high vertical percolation.
2. Many feeder roots lying close to the soil surface may be damaged or cut off while opening the furrows.

6.4.1.5 Pitcher System

Pitcher with a hole at the bottom is buried in the tree basin where feeder roots are confined. When it is filled with water, water is released slowly as droplets, which meet the water requirement of plant. Number of pitchers per tree depends upon the spread of the tree. However, 4 to 5 pitchers per tree are sufficient for the tree having 3 m spread. This system is highly a boon in the arid region where availability of water is limited.

6.4.2 Sub-Surface Irrigation

As suggested by the name, the application of water to fields in this type of irrigation system is below the ground surface so that it is supplied directly to the root zone of the plants. Such systems are generally preferred in semi-permanent/permanent installations. In this system, water is applied below the soil surface to raise the water table into or near the plant root zone. It is typically used in conjunction with subsurface drainage, or controlled drainage. Subsurface drainage lowers the water table and removes excess water through open ditches or perforated pipe. Water table depth can be controlled by installing a weir on the drainage system. During wet periods, the water table is lowered so the root zone remains unsaturated. During dry periods, water is pumped into the drainage system to raise the water table and provide additional water for plant growth. In some situations, drained water can be stored for use when irrigating.

There may be two ways by which irrigation water may be applied below ground and these are termed as: Natural sub-surface irrigation method and Artificial sub-surface irrigation method. They are discussed further below:

6.4.2.1 Natural Sub-Surface Irrigation Method

Under favourable conditions of topography and soil conditions, the water table may be close enough to the root zone of the field of crops which gets its moisture due to the upward capillary movement of water from the water table. The natural presence of the water table may not be able to supply the requisite water throughout the crop growing season. However, it may be done artificially by constructing deep channels in the field which may be filled with water at all times to ensure the presence of water table at a desired elevation below the root zone depth. Though this method of irrigation is excellent from both water distribution and labour-saving points of view, it is favourable mostly for the following conditions:

i. The soil in the root zone should be quite permeable.
ii. There should be an impermeable substratum below the water table to prevent deep percolation of water.
iii. There must be abundant supply of quality water, i.e., one which is salt free; otherwise there are chances of upward movement of these salts along with the moisture that likely to lead the conditions of salt incrustation on the surface.

6.4.2.2 Artificial Subsurface Irrigation Method

The concept of maintaining a suitable water table just below the root zone is obtained by providing perforated pipes laid in a network pattern below the soil surface at a desired depth. This method of irrigation will function only if the soil in the root zone has high horizontal permeability to permit free lateral movement of water and low vertical permeability to prevent deep percolation of water. For uniform distribution of water percolating into the soil, the pipes are required to be very closely spaced, say at about 0.5 m. Further, in order to avoid interference with cultivation the pipes have to be buried not less than about 0.4 m below the ground surface. This method of irrigation is not very popular because of the high expenses involved, unsuitable distribution of subsurface moisture in many cases and possibility of clogging of the perforation of the pipes.

Subsurface drip irrigation (SDI) is a low-pressure, high efficiency irrigation system that uses buried drip tubes or drip tape to meet crop water needs. SDI technologies have been a part of irrigated agriculture since the 1960s; with the technology advancing rapidly in the last two decades. An SDI system is flexible and can provide frequent light irrigations. This is especially suitable for arid, semi-arid, hot, and windy areas with limited water supply. Farm operations also become free of impediments that normally exist above ground with any other pressurized irrigation system. Since the water is applied below the soil surface, the effect of surface infiltration characteristics, such as crusting, saturated condition of pounding water, and potential surface runoff (including soil erosion) are eliminated during irrigation. With an appropriately sized and well-maintained SDI system, water application is highly uniform and efficient. Wetting occurs around the tube and water moves out in all directions. Figure 1 shows moisture distribution through a sub-surface drip system. Subsurface irrigation saves water and improves yields by eliminating surface water evaporation and reducing the incidence of disease and weeds. Water is applied directly to the root zone of the crop and not to the soil surface where most weed seeds winter over. As a result, germination of annual weed seed is greatly reduced, and lowers weed pressure on beneficial crops. In addition, some crops may benefit from the additional heat provided by dry surface conditions, producing more crop biomass, provided water is sufficient in the root zone. When managed properly, water and fertilizer application efficiencies are enhanced and labour needs are reduced. Field operations are also possible, even when irrigation is applied.

Fig. 1: Subsurface drip irrigation system

6.4.3 Micro-Irrigation

Slow and regular application of water directly to the root zone of plants through a network of economically designed plastic pipes and low discharge emitters is known as micro-irrigation system. Micro-irrigation can maximize crop productivity and protect the environment through conserving soil, water and fertilizer resources, while also increasing farmer income. However, a majority of smallholders in developing countries are deprived of this technology due to its high capital cost and non-adaptability to small land holdings. Until recently, it has been too expensive to be affordable for poor families and too large for tiny plots of land.

Now low-cost micro-irrigation systems have been developed. Most of the components in a typical low-cost micro-irrigation system are manufactured from polyvinylchloride and various types of polyethylene and polypropylene. The manufacturing technology is based on a simple extrusion or injection moulding process. Because of this, manufacturers of plastic pipes can easily adapt the technology to the needs of the small holders and enable them to cultivate high-value cash crops with small amounts of water to increase their income. With the use of the technology, small holders are able to increase their income up to two to three times what they make from traditional crops. With available water, farmers can also increase their productive area when using micro-irrigation System.

Under continuously favourable moisture conditions, improved cultivars can attain their high yield potential and can respond to great amounts of fertilizer and more intensive management practices. The advent of advanced irrigation system (sprinklers, drippers or tricklers, micro-sprayers, bubblers etc.) has made it possible to establish and maintain soil moisture conditions at a more nearly optimal level. Since a high frequency/micro-irrigation system can be adjusted to supply water at very nearly the exact rate required by the crop, one no longer needs to depend on the soil's own ability to store water during long interval between irrigations. The consequences of this fact are far-reaching. New lands, until recently considered very unsuited for irrigation, can now be brought into production. One outstanding example is the case of coarse sand and gravels, where moisture storage capacity is minimal and where the conveyance and spreading of water by surface flooding would involve inordinate losses by excess and non-uniform seepage. Such soils can now be irrigated quite readily even on sloping around without expensive levelling, by means of drip or micro-sprayer systems.

Advantages of Micro-Irrigation System

- ***Improved Yield:*** Slow and regular application of water and nutrients uniformly to all plants improves product quality and uniformity, and increases yield.
- ***Water Saving:*** Water saves around 50-60 %, compared to traditional irrigation methods. This means, using micro-irrigation system, a farmer can irrigate more crop area per unit of water used. So, water use efficiency is enhanced.
- ***Decrease in Labour Load:*** Less labour is required for irrigation, weeding and fertilizer application compared to traditional management systems.
- ***Fertilizer Saving:*** Fertilizer losses are minimized by adapting MI system that ultimately reduces fertilizer costs.
- ***Energy Saving:*** Most of the MIS are gravity operated systems or operated with low horsepower pumps, reducing energy demand for irrigation.
- ***Easy in irrigating on Terrain*:** MIS can be used on undulated terrain (hilly areas) where irrigation by traditional methods is difficult.
- ***Salinity:*** Due to slow and regular application of water by MIS, concentration of salts in the root zone is reduced and micro-leaching salts are kept away from the root zone.
- ***Disease control:*** Regular irrigation ensures timely inter-cultural operations and spraying, allowing better control over potential crop diseases. It also reduces the incidence of diseases common with flood irrigation.
- ***Reduced Cultivation Cost:*** Slow and regular application of water keeps an optimum soil-water-air ratio in the soil which is essential for healthy plant growth. It also reduces the need for frequent inter-culturing, weeding, etc. Combined with the above-mentioned savings, cultivation costs on the whole are reduced.
- ***Suitable to Variety of Crops:*** A number of different crops can be irrigated using MIS including vegetable crops, fruit crops, commercial cash crops, flowers, etc

6.4.3.1 Drip/Trickle Irrigation

It is the process of slow application of water in the form of discrete, continuous drops through mechanical devices called emitters or applicators located at selected points along the water delivery line.

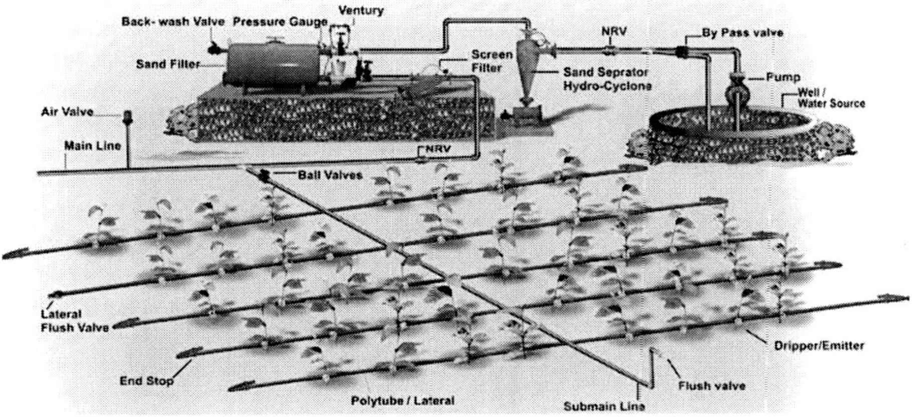

Fig. 2: Lay out plan for drip irrigation system for horticultural crops

Drip Irrigation Concept: In surface irrigation, particularly in flood irrigation system, initially the soils become over saturated that leads to filled up all air spaces in soil with water which creates anaerobic condition around root zone and ultimately hampers the root respiration. Then gradually moisture content in soil decreases due to evaporation, transpiration, percolation etc., after few days, a moisture deficient condition arises in soil and plant starts showing wilting symptoms. At that time again irrigation is given. During this irrigation cycle, plant is always in moisture stress i.e. either excess moisture stress or moisture deficit stress. However, with drip irrigation system, the soil moisture level is always maintained close to Field Capacity. This enables easy water uptake by the plants, avoids the stress to plant and maintains proper aeration around root zone. That ultimately increases the productivity. The Figure 3 below well explains the importance of drip irrigation system.

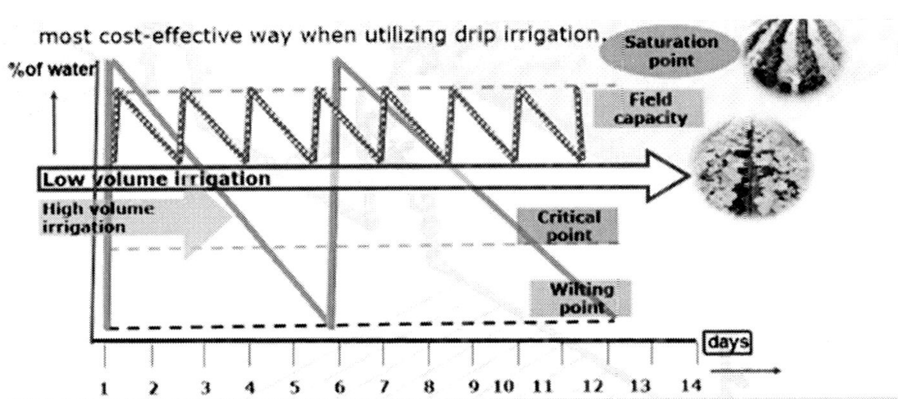

Fig. 3: Role of Drip irrigation system

Soil Wetting Pattern Under Drip Irrigation System: It depends on type of soil. In clay soil, loamy soil and sandy soil the wetting pattern through drip irrigation system is oval, round and spherical, respectively.

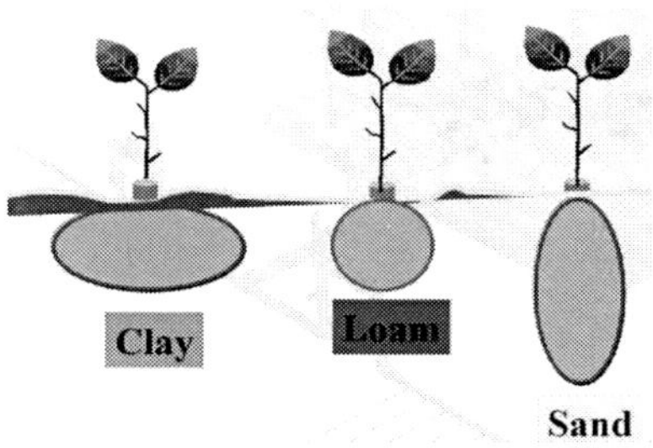

Fig. 4: Soil wetting pattern

Basic Components of Drip System

1. **Water Source:** The Drip System is a low-pressure system that uses gravity to increase water pressure. The water source can be an overhead tank placed at a minimum of one metre above ground level for smaller systems up to 400 m^2 area. For larger systems, the height of the tank should be increased. If the height of the tank is not increased, the system can be connected to a pump that lifts water from sources such as a well, farm pond, storage tank, or a stream / canal. A manually operated pressure pump also can be used to lift water from a shallow water table (up to 7 metres) and used for the system.
2. **Control Valve:** A valve made of plastic or metal to regulate required pressure and flow of water into the system. There are valves of various sizes depending on the flow rate of water in the system.
3. **Filter:** The filter ensures that clean water enters the system. There are different types of filters available such as sand filter, screen filter, hydro-cyclone filter and disc filter. Different sizes of filters are also available depending on the flow rate of water in the system. The type of filtration should be selected primarily according to the type of water source, water quality and required flow rate of the system.
    a. *Sand filter*: It stops the impurities and allows only clean water to go ahead, thus preventing clogging of the system. It is available in 30, 60, 90 litres.
    b. *Screen filter*: A screen filter is a type of filter using a rigid or flexible screen to separate sand and other fine particles out of water for irrigate.

c. *Hydro cyclone filter*: The principle of hydro cyclone filter is Centrifuge separation of solids from the Water.

d. *Disc filter*: Primary filter for bore wells and secondary filter for open water sources.

4. **Mainline:** Pipe is made of poly vinyl chloride (PVC) or polyethylene (PE) to convey water from the source to the submainline. PE pipe material is normally made from high-density polyethylene (HDPE), low-density polyethylene (LDPE) and linear low-density polyethylene (LL-DPE). The size of pipe depends on the flow rate of water in the system.
5. **Sub-mainline:** Made of PVC / HDPE / LDPE / LLDPE pipe to supply water to the lateral pipes. Lateral pipes are connected to the sub-main pipe at regular intervals. The size of pipe depends on the flow rate of water in the system.
6. **Lateral:** Pipes are made of LLDPE or LDPE placed along the rows of the crop on which emitters are connected to provide water to the plants directly. The lateral pipe size is from 12 mm to 16 mm in most drip systems.
7. **Emitters:** A device through which water is emitted at the root zone of the plant with required discharge. Different types of emitters used in drip irrigation systems are described below:

   i) *Micro-tube*: Straight or curled LLDPE tube with an inner diameter ranging from 1 to 1.2 mm. The discharge from the micro-tube is directly proportional to the operating pressure and inversely proportional to its length. The operating pressure that is required can be as low as 1-5 meters (0.43 kPa – 2.17 kPa).

   ii) *Drip Tape / Built-in Dripper*: It has built-in drippers / outlets on the lateral line which give a continuous wetting strip. It is mainly used for row crops. The operating pressure required is from 1-5 meters (0.43 kPa – 2.17 kPa).

   iii) *Drippers*: Water coming from the laterals reaches the plant root zone through drippers. Types of crop, soil and water requirements are deciding the factors for drippers spacing on the lateral.

Maintenance of Drip System

i. Regular back washing and cleaning of sand filter is essential.

ii. Frequent cleaning of emitters and drippers is necessary to avoid choking.

iii. Cleaning of main and sub-main pipes at regular interval is suggested.

v. Cleaning of PVC pipes and laterals with acid or chlorine helps to remove clogging.

Calculation of Water Requirement of Plant

Water requirement of the plant per day is equivalent to the rate of potential evapo-transpiration per day. Evapo-transpiration is the quantity of water transpired by the plants and water evaporated from the soil surface. The reference values for evapo-transpiration are normally available for a particular area at the nearest meteorological observatory.

Water requirement (WR) can be calculated as:

$$WR\left(litres\ per\ day\right) = ET \times Kc \times Cp \times Area\left(m^2\right)$$

Where, ET is evapo-transpiration (mm per day), Kc is crop factor and Cp is canopy factor.

If specific crop factor values are not available, it can be assumed as one. The canopy factor is the percentage of area covered by plant canopy (foliage). It varies according to the growth stage of the plant. The area for orchards is the multiplication of the distance from plant to plant (m) and distance from row to row (m). For row plantation, the unit area can be taken to calculate water requirement.

Example: Calculate peak water requirement for grapes planted at the spacing of 2 m by 2 m. Assume peak ET for the area as 6 mm per day, crop factor for grape 0.8 and canopy factor 0.8.

Peak water requirement per day = 6×0.8×0.8×2×2

= 15.4 litres per day per plant

The drip system has constant discharge at the given pressure. Therefore, operating time can be varied to provide the required amount of water depending on the season.

Operating Time / Irrigation Schedule

Operating (irrigation) time is the duration of irrigation system operation that provides the required amount of water for the plants. It can be calculated as follows:

$$Irrigation\,time\left(hrs / day\right) = \frac{Water\ requirement\left(litres\ per\ day\right)}{Application\,rate\left(litres\ per\ hour\right)}$$

$$Application\,rate = \frac{Q\left(discharge\ of\ dripper\ in\,lit\right)}{Inline\ to\ inline\ spacing\left(metre\right) \times emitter\ spacing\left(metre\right)}$$

Example 1

Calculate the irrigation time required for a 1000 sq. meter pineapple plantation with a daily water requirement of 800 litres and a micro-tube system discharge rate of 200 litres per hour.

$$Irrigation\,time\left(hrs / day\right) = \frac{800}{200} = 4\,hrs / day$$

Example 2

Calculate irrigation time for a papaya tree with daily water requirement of 10 litres per day per plant and provided the micro-tube system with a discharge rate of 4 litres per hour.

$$Irrigation\,time\left(hrs / day\right) = \frac{10}{4} = 2.5\,hrs / day$$

Example 3

Calculate the irrigation time required for a 4000 sq. metre (1 acre) of banana plantation with a daily water requirement of 1400 litres and dripper discharge 800 litres water per hour. The distance between the drippers is 2 m and distance between two drip line pipes is 2 m.

$$Application\,rate\left(lit / hr\right) = \frac{Q\left(discharge\,of\,dripper\,in\,lit\right)}{Inline\,to\,inline\,spacing\left(metre\right) \times emitter\,spacing\left(metre\right)} = \frac{800}{2 \times 2} = 200$$

$$Irrigation\,time\left(hrs / day\right) = \frac{1400}{200} = 7\,hrs / day$$

6.4.3.2 Sprinkler Irrigation

Over head or sprinkler irrigation refers to the application of water to crops in the form of spray from above the crop like rain. Water under pressure is carried and sprayed into the air above the crop through a system of pipes laid on the ground. Nozzles of fixed type or rotating under the pressure of water are set at suitable intervals in the distribution pipes. Water is sprayed through these perforations or nozzle over the crop wetting both the crop and soil.

In Sprinkler irrigation, two types of equipment are in use, *viz.*, micro-sprayers and micro-sprinklers. Micro-sprayers and static micro jets are non-rotating type with flow rates ranging from 20 to150 l/h, sprinkler heads will only spray in one direction. Whereas, micro-sprinklers are rotating type with flow rates ranging from 100 to 300 l/h. Emitters rotate as they spray and delivering water across a broader area. Rotating heads are often preferred because it allows for the installation of a single sprinkler array to cover a big area. A spray irrigation system, the irrigation sprinklers may be fixed in place, or located on movable frames.

A more "modern" alternative to the high-pressure water guns is the low-pressure sprinkler system. Here, water is gently sprayed downward onto plants instead of being shot high in the air. Low-pressure systems are more efficient in that much less water evaporates or is blown off the fields, if there is a strong wind present.

Advantages

1. Water is used economically and water use efficiency is high.
2. It can be adopted even on undulating topography.
3. It is adopted where water is scarce and costly.
4. Soil water can easily maintain at favourable tension for optimum growth and yield.
5. Application of fertilizer and pesticide can be possible.
6. Uniform application of water is possible.

Disadvantages

1. Initial high capital investment and high operational cost.
2. Technical person for its operation and maintenance is required.
3. Clean water is needed to avoid clogging of nozzles.
4. Areas with hot winds are unsuitable.
5. Pipe system laid on the soil surface may interface with farm operations and movement of implements.

Classification of Sprinkler System

Sprinkler irrigation system may be classified into two ways depending on

1. Types of nozzle system
    a. Nozzle line sprinkler system
    b. Rotary head sprinkler system
    c. Fixed head sprinkler system
    d. Propeller type sprinkler system
    e. Perforated pipe system
2. Portability of the system

a. *Nozzle line sprinkler system:* It consists of one or more pipes of relatively smaller diameter having a single row of fixed small nozzle spaced at uniform interval along their entire length. Pipes are supported on rows of posts at a height convenient to spray over crops and can be rotated 90°. Water is sprayed at a pressure of 2-3 atm at right angles to the pipe line and at an angle of 45° to the area of both sides and width of strip covered varies from 6-15 m according to the pressure of water and nozzle used.

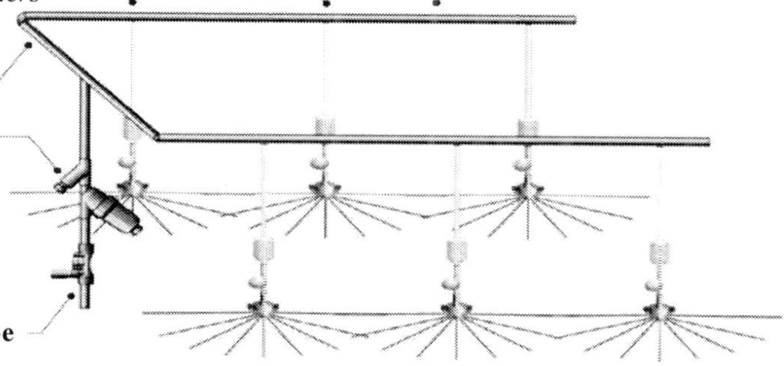

Fig. 5: Nozzle line sprinkler system

b. *Rotary head sprinkler system*: This system consists of nozzles rotate under pressure of water and spray water in a circular way. Nozzles are fitted on riser pipes attached to lateral pipelines at uniform intervals along the length of pipe. There may be a single nozzle or double nozzles on a riser pipe. Laterals are laid on the ground and spaced at about 15 m intervals. A working pressure of 1.4-3.4 atm is used for high pressure nozzles.

Fig. 6: Rotary head sprinkler system

c. *Fixed head sprinkler system:* Nozzle in this system remains stationary and spray water in one direction only to which spray nozzle is directed. The system is used extensively in orchards and nurseries. It has high water application rates. The spray is usually fine which is helpful for irrigating seedling in nursery.

Fig. 7: Fixed head sprinkler system

d. *Propeller type sprinkler system*: In this system, sprinklers are mounted on a horizontal pipeline which is held above the crop by a horizontal super structure centrally pivoted over a wing like fashion. Sprinkler pipeline with spray structure propel slowly and spray a wide range. The whole structure can be wheeled to a new position through path ways in the field. Water is conveyed to the sprinkler pipeline by a rubber hose either directly from pumping plant or from main line.

Fig. 8: Propeller type sprinkler system

e. *Perforated pipeline system*: This system includes lateral pipes perforated at regular intervals in a definite pattern to spray water through these perforations. These pipes are installed in rows at an interval of 6-15 m and working pressure is only from 0.3-1.0 atm. An overhead tank suffices the need to create the pressure. Pipes are perforated to spray the area on both sides of a pipe and strip of 6-15 m wide. The system is adapted for irrigation of lawns, gardens and small vegetable field.

Fig. 9: Perforated pipeline system

2. Portability of the system: On the basis of portability of sprinkler units they are classified as a) Portable, b) Semi-portable and c) Stationary. Portable and semi-portable types are mostly used in agriculture. These systems can be removed manually or mechanically.

Basic Components of Sprinkler Systems: The basic components are almost same as drip irrigation system, i.e., water source, pump, filter, main and sub-main line, lateral. The only difference is at their head. Sprinkler head is mainly three types as follows:

i) **Micro-Sprinkler:** It has a small rotating device to spray water as light precipitation. It covers an area with radius of 3-4 meters. Operating pressure required is 5-10 meters.

ii) **Mini Sprinkler:** It has a small rotating device to spray water as light precipitation. It covers an area with a radius of 6-8 meters. The operating pressure required is 5-15 meters.

iii) **Impact Sprinkler:** It is made of metal or plastic and has a spring which makes the hammer move, rotating the sprinkler. It covers an area with a radius of 10-12 meters and operating pressure required is10-20 meters.

Micro Sprinkler

Mini Sprinkler

Impact Sprinkler

Efficiency of Sprinkler Irrigation:

1. Sprinkler irrigation is more efficient than surface irrigation.
2. Evaporation losses from sprinklers depend on the RH, temperature, wind velocity and fineness of drops that in turn depends on the water pressure and nozzle size. It may be only 2-8 % of the total sprinkler discharge.
3. Leaf transpiration is greatly reduced due to free water on the leaf surface and high humidity near the leaf surface.
4. The water application efficiency is high and it is about 85-90%.

Principles of Selecting Sprinkler System

1. Water requirement of crop.
2. Capacity of the system to apply water equal or less than intake capacity of the soil.
3. The system with maximum water application efficiency.
4. Cost efficient from the point of crop production economics.
5. Soil texture.
6. Cost and availability of water.

Spacing of Sprinklers and Laterals

The uniformity of water distribution from sprinkler depends on the pressure of wind velocity, rotation of sprinklers, spacing and so on. Generally, a higher amount of water is applied near the sprinkler and the amount decreases gradually with distance from the sprinkler. Sprinklers are arranged along a lateral not more than 50% of the diameter of the coverage by an individual sprinkler. The distance between successive positions of laterals should not exceed 65% of the diameter of coverage by an individual sprinkler.

References

Aquastat database, Food and Agriculture Organization of the United Nations (FAO). www.fao.org/nr/water/aquastat/main/index.stm.

Bjorneberg DL (2013). Irrigation Methods, Reference Module in Earth Systems and Environmental Sciences. Elsevier. Pp.1-10

Dragoni D, Lakso AN and Piccioni RM (2005). Transpiration of apple trees in a humid climate using heat pulse sap flow gauges calibrated with whole-canopy gas exchange chambers. Agricultural and Forestry Meteorology, 130, pp. 85-94.

Fangmeier DD, Clemmens AJ, El-Ansary MY, Strelkoff TS and Osman HEB (1999). Influence of land-levelling precision on level-basin advance and performance. *Transactions of the ASAE,* **42**: 1019–1025.

Rana G and Katerji N (2000). Measurement and estimation of actual evapotranspiration in the field under Mediterranean climate: a review. *European Journal of Agronomy*, **13** (126): 125–153.

Xiloyannis C, (1992). Irrigazione. In: F. Lalatta, F. (Ed), Frutticoltura Generale, Roma, REDA press, pp. 597-623.

Xiloyannis C, Massai R, Piccotino D, Baroni G and Bovo M (1993). Method and technique of irrigation in relation to root system characteristics in fruit growing. *Acta Horticulturae*, **335**: 505-510.

Xiloyannis C, Montanaro G and Dichio B (2012). Irrigation in Mediterranean Fruit Tree Orchards. Chapter 15. https://www.researchgate.net/publication/224829388.

Outcomes Assessment

Part A

Answer the following questions true or false.

1. Broadly, there are five main methods to apply water to the plants. True/False
2. Sprinkler irrigation is equally efficient to surface irrigation. True/False
3. Evaporation losses from sprinklers depend on the Relative Humidity. True/False
4. Filter ensures that clean water enters the system. True/False
5. Regular back washing and cleaning of sand filter is not required. True/False

Part B

Answer the following questions.

1. What is micro-irrigation?
2. Water requirement of plant per day is equivalent to the rate of potential — per day.
3. Define irrigation time.

4. What is drip irrigation?
5. Surface irrigation has five/seven main techniques.

Part C

Write a brief note on each of the following.

1. Basic components of drip system.
2. Sub-surface irrigation.
3. Basin irrigation.
4. Sprinkler irrigation.
5. Soil water-holding capacity.

7

Soil Management

7.1 Introduction

Soil is an essential component of ecosystem sustaining all living vegetation, microorganisms, herbivores and carnivores. It is the outermost layer of the earth. It varies greatly in its physical and chemical properties which depend on its age and on the conditions (parent material, climate, topography and vegetation) under which it was formed. Processes such as leaching, weathering and microbial activity combine to make a whole range of different soil types, each of which has particular strengths and weaknesses for agricultural/ horticultural production.

The materials present in the soil are mostly in three states, i.e., solid, liquid and gaseous. They differ in their percentage from soil to soil. A good soil should have 5% organic matter, 45% mineral matter, 25% air and 25% water. So, while establishing an orchard, whole soil profile should be taken into consideration. Digging a pit shows the different layers of soil profile. Usually these are same and not distinct in many areas. Inorganic materials are the major component of most soils. They consist largely of mineral particles with specific physical and chemical properties which vary depending on the parent material and conditions under which the soil was formed. It is the inorganic fraction of soils which determines soil physical properties such as texture and has a large effect on structure, density, water retention and aeration. Air and water percentage in soil vary significantly with soil texture and weather conditions. The soil texture is a property which is determined largely by the relative proportions of sand, silt and clay particles content in it. The classification of soil based on particle size is termed as textural classification of soil. As per USDA classifications, clay particle size is less than 0.002 mm, silt size varies in between 0.002 to 0.05 mm and sand size in between 0.05 to 2.0 mm. On the basis of percentage of these particles present in a soil, the type of soil can be easily determined. Accordingly, soils are classified as sandy loam, silt loam, clay loam and sand. A sandy loam soil may consist of 75% sand, 14% silt and 11% clay particle. A clay loam shall have 27% sand, 34% silt and 39% clay.

Soil organic matter contributes to stable soil aggregates by binding soil particles together of various shape and sizes. Maintaining the soil structure in fine granular form is very important. Organic matter though present in small amount (0.5% - 5%) in soil but is the most important component. It greatly improves the soil water holding capacity, but decreases the plasticity and cohesion. Soil organic matters strongly influence the population of soil microorganism. The organic matter fraction of soil come from the decomposition of animal or plant products such as faeces, roots, leaves etc. Decomposition of this organic matter by microbial activity releases nutrients for the growth of other plants. Organic matter content of orchard soil can be improved by growing green manuring (sunhemp, dhaincha, cowpea, guar, berseem, arhar etc.) and/or leguminous crop as intercrop during juvenile period of fruit plants. It will maintain the optimum soil organic matter and improve the soil health as well as improve the microbial activity in soil which will be mutually benefitted.

Soil microorganisms are the primary agents responsible for the decomposition of organic matter. The soil contains numerous organisms ranging from microscopic bacteria to large soil animals such as earthworms. The soil microorganisms include bacteria, fungi, actinomycetes, algae, protozoa and nematodes. The diversity of soil organisms can both assist and hinder plant growth. Beneficial activities include organic matter decomposition, nitrogen fixation, transformation of essential elements from one form to another, improvement in soil structure through soil aggregation, and improved drainage and aeration. However, under some circumstances soil organisms compete with plants for nutrients.

Large active microbial communities are important to reduce pest and pathogen pressure and maintain root health. They do this both by competing for resources and by directly attacking pests and pathogens (van and Ginkel 2001; Garbeva *et al.* 2004). For example, some groups of microbes produce antibiotics (*Streptomyces*, and numerous strains of *Micromonespora* and *Nocardia*). Some *Pseudomonas* strains produce antibiosis reducing pathogen pressure. Other microbes such as the fungi *Trichoderma,* actively parasitize plant pathogens. Orchard soils higher in *Trichoderma*, as well as Streptomyces and *Pseudomonas* bacteria, have been found to be more suppressive to the pathogens involved in replant disease than soils with lower levels of beneficial (Weerakoon *et al.* 2012).

Bacteria are the smallest and most numerous microorganisms present in the soil. They make an important contribution to organic matter decomposition, nitrogen fixation and the transformation of nitrogen and sulphur and make available for plant. The fungi and actinomycetes contribute beneficially to organic matter decomposition. The group of large soil animals include earthworms, which incorporate organic matter into the soil as well as improving aeration and drainage by means of their channels. But some soil fungi, nematodes and insects feed on roots and lateral shoots which is detriment to plants.

Soil earthworm count is more in orchards where FYM and green manuring is done. Fruit trees like mango, guava, sapota, ber, aonla etc. are deep rooted trees and having larger root system for searching of water and nutrient for its growth. However, many roots get damaged and die due course of time. These dead roots contribute continuously for microbial activity. Due to this orchard soil has more earthworm than field crop soil. However, earthworm population reduces in soil where weedicide is used regularly.

Performance of orchard soil is greatly inclined on soil reaction that is soil pH. Depending on the soil reaction, it may be acidic or alkaline or neutral in nature. Accordingly, kind of fruit plant should be selected for planting otherwise soil may be reclaimed. Soil acidity is common where sufficient leaching of exchangeable base from surface layer of soil has taken place. Alkaline condition occurs in the soil where bases have accumulated. Alkaline soils are prevalent in central districts and irrigated zone of Punjab. Soil becomes acidic, if in a given soil clay micelle, the adsorbed hydrogen is higher, which results into higher H^+ ions concentration in the soil solution. The alkaline soil is results of hydrolysis of soil colloids saturated with calcium, magnesium and potassium cations. The concentration of H^+ ion in the soil solution decreases and simultaneously OH^- ions concentration will increase. Ultimately the soil solution will turn alkaline in reaction. The pH of the solution depends on their relative amount of adsorbed hydrogen ion and adsorbed metallic cations. The pH of the soil does not remain constant. Rain, irrigation water, addition of manures and fertilizers affect the soil pH. Most of the soil in arid irrigated areas of North Indian plains has slightly to moderately high soil pH (alkaline). These soils which are both calcareous and strongly alkaline can be put under ber, guava, aonla and pomegranate fruit cultivation (Sharma and Singh, 2011).

The effect of orchard soil management practices such as cultivation, zero-tillage with the use of herbicides, mulching and cover cropping, on the growth of deciduous fruit trees is reviewed. Emphasis is placed on the mechanisms and processes by which the different management practices affect tree growth through the alteration of various components of the orchard environment including nutrient availability, soil physical properties, moisture availability, abundance of soil flora and fauna, and prevalence of pests and diseases. The relative merits of each system are discussed below and each management practice has its own advantages and drawbacks both from the scientific and practical standpoints.

7.2 Orchard Soil Management

Soil provides the structural support and is the source of water and nutrients for plants used in agriculture/ Horticulture. Productivity of soil is the fertility of the soil that produces crops with a set of management practices. Most of the soil management practices in the orchard enhance nutrient and water supply needed for growth and production of fruit trees. This they do so by improving soil nutrient status, conserving soil moisture, altering of soil properties like soil pH and soil structure, and maintaining soil organic matter contents and useful soil microorganisms. There is no single management practice that can be followed for all types of soil in fruit plantation. As per the crop requirement, climate and soil conditions, different management practices are followed.

7.2.1 Objectives of Orchard Soil Management

1. To improve soil organic matter.
2. To improve availability of nutrients in soil.
3. To preserve soil structure.
4. To improve irrigation facility to fruit plants.
5. To control weeds in orchard.
6. To facilitate easy movement in the orchard.
7. To save fruit plants from frost and freezes.
8. To reduce the root diseases of plants.
9. To maintain optimum air and water percentage in soil.
10. To improve the soil beneficial microorganism activity.
11. To maintain the soil pH as per need of the particular fruit tree plantation.

7.2.2 Soil Management Practices

There are mainly five soil management practices that help in achieving the above said objectives. These practices are: (a) Permanent or Partial Sod culture, (b) Cover crops, (c) Clean cultivation, (d) Mulching and (e) Weedicide.

7.2.2.1 Permanent or Partial Sod culture

In sod culture a thick carpet of grass is allowed to grow to cover the vacant area that lies in the orchard. Occasionally the grass is mowed in order to increase the organic matter content of the orchard soil and to decrease competition for moisture and nutrients. The practice of establishing a complete sod cover on the orchard floor often causes a reduction in vigour and growth of the fruit plantations, which is commonly associated with reduced nutrient uptake, particularly the nitrogen. Fruit trees under permanent sod usually have lesser leaf nitrogen contents, particularly during mid to late growing season (Sharma 2012). Due to shade, permanent sod culture cannot survive during winter months. The poor growth of sod is beneficial during summers, as sod do not compete for water and nutrients with the fruit plants. Permanent sod is also very difficult to maintain in orchards, where flooding or segment irrigation is followed. Sod checks the flow of water and huge quantity of water is wasted through percolation. Sod cannot be created in orchards which are under sprinkler or drip irrigated. These irrigation systems do not irrigate entire vacant area; hence the sod species shall die as a result of drought. Grasses and weeds do grow in orchards particularly in North Indian plains. These make lush growth during monsoons. Although weeds and grasses need to be removed from fruit tree basins to save water and nutrients, even then these do add some quantity of organic matter to the soil.

Soil under permanent grass may begin to lose water 15-30 days earlier in the spring as compared to the cultivated soil. Grasses in orchard require a large proportion of water for maintaining growth, and therefore in non-irrigated orchards this competition of water with fruit trees can become critical. The extent of soil moisture depletion varies with the vigour, rooting depth, and frequency of mowing of the orchard sod. Shallow rooted grasses such as Kentucky bluegrass (*Poa pratenses* L.) or annual blue grass (*Poa annua* L.) deplete less moisture from the orchard soil profile than deep rooted sods such as Landino clover (*Trifolium repens* L.) or perennial ryegrass (*Lolium perenne* L.) (Sharma 2012). The Khabal grass should be allowed to grow on the paths, segments and vacant space outside the tree basins to form a sod. This allows rain water percolation and checks surface runoff. It also lowers the severe hot summer temperature of the orchard and lessens soil compaction.

Sod culture is not practiced in the younger plantations due to growing of intercrops for early three to four years of planting an orchard. In later years if sod culture is applied in orchard soil it will be beneficial in checking surface water runoff and shall protect top soil from wind and water erosion. Plant root holds the soil particles in place. In Europe, America and Australia sod culture is generally practiced in orchards. It will be useful in orchards planted in hilly areas on contours.

During summer season, regular mowing of grasses should be done to conserve moisture in soil by reducing water use by grasses. However, sod improves other physical properties such as increase in soil porosity and decrease in bulk density. Decrease in bulk density means an increase in total pore space available for root growth and increase in water holding capacity. Such changes are likely to increase earthworm populations and also expected to reduce surface moisture runoff and erosion and moisture conservation.

Interestingly, nitrogen return to the orchard floor by sod mowing is estimated to be higher at 400-600 kg per hectare and overall total soil nitrogen under sod is higher than over herbicides and clean cultivation. Sod cover also supplies high amount of available nitrogen; however, it also uses large amount of the available nitrogen and thus decreases its availability to the tree roots. On the contrary, sod culture usually increases leaf phosphorous concentration in apples. Elevated leaf phosphorus concentration may be due to an inverse relationship between phosphorus and nitrogen uptake or more mobility of phosphorus induced by mycorrhizal infection transferable from cover crops to the tree roots. Sod can also increase leaf potassium concentration and reduce leaf scorch symptoms typical of potassium deficiency. Potassium returns are high from sod clippings with annual estimates of 600 kg per hectare, for orchards under long term sod culture. Sod cover may increase soil and leaf calcium and magnesium contents and decrease leaching losses from the soil. However, if magnesium supply is inadequate, it may be associated with a sod induced increase in leaf potassium. Sod may also decrease iron deficiency symptoms as observed elsewhere in apple (Sharma 2012).

Very few orchards in India carry out sod culture, whereas in Australia many citrus orchards are under permanent sod culture. In India too some kind of adjustment between clean cultivation and sod culture shall be advantageous.

7.2.2.2 Cover Crops

Growing of cover crops in orchard is the best system of soil management in hands of the farmer. Many kinds of crops can be grown in orchards as cover crops. It is the cheapest and most effective method of maintaining soil productivity and soil structure. The orchards where some kinds of cover crop (intercrop) have been grown, the soil is open and friable due to the addition of organic matter and root penetration. Thus, the water holding and nutrient holding capacity is improved along with increased soil aeration. In arid irrigated zone of north India, it is beneficial because the soils are sandy and improvement of organic matter in such soils is advantageous to the orchards. Especially water remains in the upper root zone near the surface soil, which is useful to the newly planted plants with shallow roots. In heavy soils the root activity of the cover crops and organic matter addition help in breaking the compact clay soils, the infiltration of irrigation water is also improved. With the improvement in soil structure, fruit tree roots can spread and penetrate deep, subsequently manage to collect more water and nutrients for the growth of fruit trees. The profuse growth of the fruit trees leads to higher yield than the normal.

Legume cover crops like cowpeas, peas, beans, clovers and green manuring crops serve as a better cover crop. It adds nitrogen in proportion to the amount of top growth. Arhar can be grown in sandy areas. It will protect the soil from wind erosion. Cereals like oats, barley, wheat, rye, and sarson can be grown as cover crops. The cereals may not fix nitrogen but due to their vigorous root system it is a good alternative to legumes. The cereals are less prone to insect attack. Cereals produce more dry matter than many legumes. The growth of a cover crop should be checked before the active growth in fruit trees begins. If the growth is not curtailed the cover crop will compete for water and nutrients with the fruit trees. The crop should be either slashed down to leave the cuttings on the soil surface. Some herbicides can be sprayed. It can be mechanically ploughed/disc, and the cover crop is left on the surface of soil. The prime objective of cover crops is to keep the cover on the soil. Due to economic reasons, this is rarely done in India. In north India, orchardists grow inter-crops in orchards.

One of the limiting factors of cover crops in vine yards is the competition for minerals and water availability where the management is inadequate. These phenomena determine deep modifications of the nitrogen content of the wine with effects on the alcoholic fermentation. To solve this problem, different mixes of cover crops, including leguminous species that supply nitrogen, should be evaluated in different areas (Benites *et al.* 2004).

Inter Cropping: During the pre-bearing period (juvenile period) farmers usually want to supplement their income by growing some crops. Most of the grower's option is for wheat and fodder crops. A few grow vegetables and very few opt for pulses and legumes. Experience has shown that if intercrops are not grown judiciously, they cause great harm to the fruit plants. Mostly, farmers revert to cereal crops as the fruit trees die slowly year after year.

At present most of the field work is done with tractor implements. The plant trunks and branches get injured by the implements, which latter on become cause of plant death. Frequent disc ploughing is very harmful to the fruit plants. It damages the feeder root system of fruit plants. Discs also make segments of various weed roots. The segmentation of baru grass (*Sorghum halepense*) roots and shoots of Daryai booti (*Cleome strica*) grass give rise to increased density of these weeds year after year. Injury caused by spade hoeing of tree basins can also become fatal to the fruit plants in early stages of their growth. Although cleaning of basins of weeds and mixing of manures and fertilizers is very necessary for the health of the fruit trees, even then trunk and root injuries become dangerous. Judicious use of implements in the orchards is very necessary. Growing of crops at the cost of plant health is not advisable. Potato growing has been very harmful to the fruit plants, particularly in kinnow and peaches. Fodders like berseem and jowar are equally harmful to fruit plants. Many orchards of patharnakh pear declined due to root rot, where berseem had been grown for a number of years. These crops should not be sown in orchards. Short duration crops and pulses are beneficial to the fruit crops and can be grown successfully in orchards repeatedly. If wheat is grown as intercrop which is usually grown throughout Haryana and Punjab and in some areas in Uttar Pradesh, a separate irrigation system for fruit plants should be prepared. This is very necessary because during April no irrigation is applied to wheat and long interval of non-supply of irrigation to the fruit plants shall adversely affect the growth and development. In addition to generate income from the cereals as intercrops, these crops add some organic matter through the decay of roots and trash.

Crops like wheat, oat, grain, oilseeds and radish (for seed) may not be grown as intercrops in grown-up litchi orchard. This is mainly because the period of seed production and subsequent development of seed in the intercrop generally coincides with the period of active fruit development and during this period the intercrop extracts maximum amount of moisture and nutrients from the soil. Perennial or exhaustive crops should be discouraged as an intercrop in the orchard. This may have devitalizing effect on the growing trees. For example, sugarcane, pigeonpea, maize, *jowar* should invariably be excluded from an intercropping programme in the litchi orchard. Vegetables that have their roots within 25 cm depth of the soil are considered good for intercropping, e.g., tomato, onion, cauliflower, beans, radish, *palak,* etc.

Non-growing of intercrops can help in the production of good quality fruits for longer years than otherwise. Pear root rot, citrus root rot and decline of peaches can only be checked by avoiding intercropping. One of the major reasons of orchard failures is growing of intercrop year after year. Growers are advised to grow intercrops judiciously only for the first two to three years after planting an orchard. Irrigational and nutritional needs of fruit tree should be met separately as per needs of the fruit plants. So, it is always beneficial in the long run not to grow intercrops in orchards. Proper care of the fruit plants will add more growth which will in-turn reduce the juvenile period by one to two years depending upon the fruit crop. With good care, kinnow comes into bearing in third year, mango in fourth year and pear and litchi in fifth year after their planting. Thus, farmers can earn more profits from fruit trees by just avoiding growing of intercrops. Fillers as fruit plants like banana or papaya in mango and litchi orchard can supplement more income than intercrops.

Cover crops provide the following additional benefits

i. Soil become easier to work.

ii. Soil does not become boggy when wet, hard and dry.

iii. Harbour many predators.

iv. The activity of earthworms is increased.

v. The dust in orchards is reduced.
vi. Weeds remain under check.

There are also some problems like poor growth of fruit tree, water and nutrient management become difficult in growing of cover crops in the orchards.

Poor Growth: Due to shade of the orchard trees, the seed germination of intercrop is affected. Due to limited solar radiation yield may decrease. For this, shade loving plants may be grown as intercrop, i.e., pineapple can be cultivated as intercrop in mango orchard. As pineapple is a partial shade loving plant, bumper yield can be expected.

Irrigation: Additional irrigations are required to meet the need of cover crops. Irrigation systems are prepared/designed to meet the water requirements of the fruit trees, not for a cover crop. In summer months fruit trees can starve of water. Frequent irrigations are needed for the cover crop for its establishment. This cannot be carried out through modified basin or basin systems of irrigation. Both main crop and cover crop have different amount of water requirement which creates a problem in water management and choosing a right irrigation system.

Nutrition: Similarly, additional fertilizer is required when intercrop or cover crop is grown with fruit tree. The nutritional requirement of main crop and sub-crop is also different which again create difficulty in nutrient management of orchard.

7.2.2.3 Clean Cultivation

Some farmers prefer clean orchards throughout the year than growing of legumes or intercrops. They are more interested in weed control in the orchards than the health of fruit trees. In this process constant ploughing or sprays of weedicides damage the root system. The feeder roots get killed which lead to nutrient deficiencies and root rot, finally early tree decline sets in. Clean cultivation damages soil structure and soil become devoid of organic matter. It is a costly practice to keep the orchards free of weeds, throughout the year. Regular cultivation destroys feeder rot system of the fruit trees from the top 10 cm of soil. In most of the orchard soils the top layer is very fertile and favourable layer for feeder root system to flourish.

Some orchardists feel that frequent deep cultivation will increase water infiltration to the deeper layer of soil to be used by the fruit plants, but in the long term it is detrimental. The loose surface soil created by frequent cultivation shall form a solid impermeable layer when irrigated. Soil microbes which continually work to break down organic matter, shall accelerate the break down due to cultivation, leading to decline in organic matter level. Frequent cultivation is a short-term remedy for improving water infiltration. The only answer for water conservation is to grow green manuring and cover crops and addition of gypsum. Excessive cultivation in peach, plum and kinnow orchards has resulted into early decline. It also results into root suckering on damaged roots particularly in pear plantations propagated on pear root suckers (*Pyrus calleryana*). These root suckers become a nuisance and are difficult to control once these starts appearing. Accordingly, the frequent cultivation of orchard soil is not recommended.

Clean cultivation to a limited extent can be useful. It improves the soil aeration by pulverizing the top soil. The nutrients which would otherwise have been utilized by weeds can be saved for fruit plants. Hoeing breaks the upper soil crust, which checks water loss through capillaries from the orchard soil. Cultivation accelerate the mineralization of organic matter and increase NO_3-N availability, however total nitrogen tends to decline under cultivation. There is always seasonal variation in NO_3-N, which is associated with the extent and timing of weed and cover re-growth. Cultivation improves leaf size and colour, leaf nitrogen concentration and availability of soil nitrogen. This may be due to mineralizing of nitrogen from the soil. However, cultivation may increase leaf nitrogen in the short term, but in the long term, it may not consistently increase leaf nitrogen. Cultivation may not have a major role in affecting the soil contents of other nutrients, however it affects leaf concentration of some of these. Different experimental reports suggest that leaf phosphorus and potassium increase, and leaf calcium concentration decreases under cultivation (Sharma 2012).

In comparison with grass, cultivation led to decreased bulk density, penetration resistance, water aggregate stability, tree root density and markedly reduced earthworm density, but also to increased total porosity, macro-porosity, saturated hydraulic conductivity, infiltration rate, and oxygen diffusion rate near the soil surface (Haynes 1981).

7.2.2.4 Mulching

Mulching is the process of covering soil around plant root area with a view to insulate the plant and its roots from the effects of extreme temperature fluctuations and to avoid weed germination and weed growth with some extraneous matter of organic or inorganic in nature. Organic mulches like dried leaves, hay, straw, saw dust etc. and inorganic material like plastic mulches are spread around the tree basin to cover the soil surface.

Inorganic mulches are usually preferred in fruit orchards located in the high hill areas having cooler climate. It also helps in raising the soil temperature in summer, required for greater nutrient absorption and roots metabolic activities. However, inorganic mulches lack soil improving properties as organic mulches.

Organic mulches add significant amount of nitrogen to the soil and consequently improve soil fertility, which may cause long term beneficial effect to orchard nutrition. Decomposable mulches increase soil organic carbon contents and soil with a favourable C : N ratio is expected to increase leaf nitrogen concentration as a result of increase in available soil nitrogen. Available phosphorus generally increases in orchard soil under decomposable mulches, especially seaweeds. Increased soil potassium availability as a consequence of mulching with hay and straw is generally noticed. This reflects large amount of potassium is added to the soil by many mulching materials; alternatively, improved potassium availability may also be due to reduced potassium fixation. Whatever the reason may be, more leaf potassium concentration as a consequence of organic mulching is most frequently observed. However, increased leaf potassium under mulching adversely affects leaf calcium and magnesium concentration, despite of increase in soil calcium and magnesium concentration. Magnesium deficiency in apple is more pronounced under mulch than sod. Organic nutrient content of soil and the kind of mulching material are also important consideration in interpreting such contradictory results. Conversely, mulching increases leaf boron concentration and thus may help to combat its deficiency (Sharma 2012).

It is well recognized that conservation of soil moisture is the most significant advantage of mulching. This useful effect is more pronounced during the drying periods of May-June. The general improvement in soil moisture is likely a consequence of both improved infiltration capacity, particularly under hay, straw or pine needle mulch and reduced evaporation. However, the moisture conservation properties of organic mulches on a silt-clay soil may sometimes be undesirable; as such soils are likely to become susceptible to water logging in winter.

Advantages

1. Conserves soil moisture by reducing evaporative losses from soil.
2. Organic mulches add organic matter and nutrients to the soil upon decomposition and thus save fertilizer requirements.
3. Organic mulches improve the soil structure, nutrient availability and micro-flora.
4. Controls soil erosion.
5. Organic mulches improve water infiltration and control weed growth due to etiolation.
6. Reduces soil temperature fluctuation.

7. Mulch acts as a cushion for dropping fruits and thus avoiding physical injury.
8. Growth, production, colour and quality of fruits are improved.

Additional benefits of organic mulches

1. Apart from the advantages listed above, organic mulches also improve root growth of the tree.
2. Increase the infiltration of water, and also improve the water holding capacity of the soil.
3. Organic mulch provides an ideal environment for earthworms and other beneficial soil organisms.
4. Organic mulches are more useful for warmer climates, i.e., valley areas and low and mid-hills areas of the Himalayan region, as these keeps the soil temperature cooler by 2-3° C in summer.

Disadvantages

1. Dry grass, hay or pine needle mulches may invite fire hazard.
2. Organic mulch may harbour insects, diseases and rodents.
3. Inorganic mulch is costly.

7.2.2.5 Weedicides (Herbicides)

Herbicide treatment resulted in a natural compaction of the surface soil with increased bulk density, penetration resistance, aggregate stability, and root density, but reduced total porosity, macro-porosity, infiltration rate and earthworm density (Haynes 1981). Soils in orchard under herbicide weed control are richer in NO_3-N in comparison to grasses orchards, and its value may be as high as 21 mg per cm^3 in the former compared to 10 mg per cm^3 of soil in the later during the mid-summer. Effect of herbicides on leaf nitrogen content is variable. It may increase leaf nitrogen temporarily, due to removal of competition or may not affect leaf nitrogen concentration consistently. Among different herbicides, Simazine might affect nitrogen nutrition directly as it may increase leaf nitrogen concentration, but only in certain growing condition. Amitrole plus Simazine herbicides can partially substitute nitrogen. Soils of untilled herbicidal strip accumulate greater quantities of phosphorus than under grassed alley, probably from the applied phosphorus. However, leaf phosphorous concentration is usually lower in trees under herbicidal strips than under grass. Potassium concentration normally does not vary considerably in the soils under herbicidal plots in comparison to phosphorus. However, vegetation control with herbicides generally causes a reduction of fruit trees leaf potassium concentration, which is occasionally accompanied by an increase in leaf magnesium. Surface soils under herbicidal plot usually are much lower in exchangeable calcium and magnesium. This is related to the cycling of these cations by orchard floor vegetation since it is known that vegetation reduces leaching of these cations from the surface soil. Leaf calcium concentration however, varies with the type of herbicide. Higher leaf calcium concentration can be observed with Paraquat treatment, whereas a combination of Paraquat plus simazine has little effect. In general, herbicide-treated orchard soils have a lower soil moisture deficit in comparison to grassed or cover cropped soil (Sharma 2012).

References

Benites J, Pisante M and Stagnari F (2004). The role and importance of integrated soil and water management for orchard development. *Proceedings of the International Seminar. FAO Land and Water Bulletin* 10, pp. 39.

Garbeva P, Veen J. A. V. and Elsas JDV (2004). Microbial diversity in soil. Selection of microbial populations by plant and soil type and implications for disease suppressiveness. *Annu. Rev. Phytopathology.*, **42**: 243-270.

doi: 10.1146/annurev.phyto.42.012604.135455.

Haynes RJ (1981). Effects of soil management practices on soil physical properties, earthworm population and tree root distribution in a commercial apple orchard. *Soil and Tillage Research*, **1**: 269-280.

Sharma KK and Singh NP (2011). Soil Management. *Soil and Orchard Management.* Daya Publication House, pp1-18.

Sharma N (2012). *Orchard Soil Management.* E-course. YSPUHF. http://ecoursesonline.iasri.res.in/mod/page/view.php?id=103186

Van Os GJ and Ginkel JHV (2001). Suppression of Pythium root rot in bulbous Iris in relation to biomass and activity of the soil microflora. *Soil Biol. Biochem.*, **33** (11):1447-1454. doi: 10.1016/s0038-0717(01)00053-0.

Weerakoon DMN, Reardon CL, Paulitz ADITC and Mazzola M (2012). Long term suppression of *Pythium abappressorium* induced by *Brassica juncea* seed meal amendment is biologically mediated. *Soil Biol. Biochem.*, **51**: 44-52. doi: 10.1016/j.soilbio.2012.03.027.

Outcomes Assessment

Part A

Answer the following questions true or false.

1. Citrus root rot can only be checked by avoiding intercropping. True/False
2. Organic mulches do not improve the soil structure. True/False
3. Organic mulches do not add significant amount of nitrogen to the soil. True/False
4. Mulching is the process of covering soil around plant root area. True/False
5. Sod culture is not practiced in the younger plantations. True/False

Part B

Answer the following questions.

1. What is clean cultivation?
2. Sod culture usually increases/decreases leaf phosphorous concentration in apples.
3. Organic mulch may harbour insects, — and —.
4. Why are growers advised to grow intercrops judiciously in an orchard?
5. There are mainly five/six soil management practices.

Part C

Write a brief note on each of the following.

1. Objectives of orchard soil management.
2. Cover crops.
3. Mulching.
4. Clean cultivation
5. Permanent or partial sod culture.

8

Fruitfulness and Unfruitfulness

8.1 Introduction

Fruitfulness refers to the state where a plant is not only capable of flowering and bearing fruit but also takes these fruits to maturity. The inability to do so is known as 'unfruitfulness' or 'barrenness'. Unfruitfulness is a major problem in many fruit crops and their varieties. It results in huge loss to growers and makes fruit cultivation less profitable.

8.2 Factors Affecting Unfruitfulness

Unfruitfulness can be due to lack of balance between vegetative growth and fruiting, lack of flowering and poor fruit set, which is result of the unfavourable environment. It can also be due to heavy cropping, leading to inhibition of fruit bud production and poor crop in the following year. Sterility also leads to unfruitfulness due to impotence, incompatibility or the abortion of embryo. This unfruitfulness is one of the serious problems of many orchards and its causes need to be understood properly for effective control and obtaining economically acceptable production level. The causes of unfruitfulness can be broadly grouped into two categories: internal factors and external factors.

8.2.1 Internal Factors Associated with Unfruitfulness

There are a number of internal factors which are associated with unfruitfulness or sterility. They have further been categorized into three major categories such as evolutionary tendencies, genetic influence and physiological factors.

8.2.1.1 Evolutionary Tendencies

The evolutionary factors leading to unfruitfulness are:

1. **Imperfect/defective flowers:** A perfect flower possesses both male (stamens) and female (pistil) parts, whereas an imperfect flower may be either staminate (functional male having only stamens) or pistillate (functional female having only a pistil or pistils). If staminate and pistillate flowers are borne on the same plant but in different locations, the species is termed as 'monoecious' for example, walnut, pecan nut,

cashew nut and chestnut. If staminate and pistillate flowers occur on different plants, the species is termed 'dioecious' for example, pistachio nut, papaya, kiwifruit, fig and mulberry. One can see the ramifications for pollination and orchard design, in that a dioecious species such as pistachio or kiwifruit must be planted in an orchard with male plants near females for pollination. Aside pollination, the male plants are useless since they do not possess ovaries that will ripen into fruit. Generally, in monoecious fruit plants, there is no or very little problem of pollination, fruit setting and fruitfulness. A number of sex forms have been reported in papaya. The evolution of unisexual flowers cuts down chances of self-fertilization completely. Of the unisexual plants, the pure staminate flowering ones are essentially barren.

2. **Heterostyly:** The presence of short style with long filaments (Thrum) for example, sapota and pomegranate, or long style and short filaments (Pin type) for example, almond and carambola is dimorphism, a type of heterostyly. The occurrence of flowers with variable length of styles is common in *Prunus* fruit crops (Suranyi 1976). Three types of flowers, namely thrum, homostylous and pin, were found in pomegranate cultivars Ganesh and Kandhari (Singh *et al.* 2006).
3. **Structural peculiarities:** When stigmatic receptivity period does not coincide with pollen viability in monoecious plants, it is known as 'dichogamy'. In dichogamy, self-pollination is prevented in perfect flowered plants, due to maturity of two sex organs at different times. If the stamens are ripened before the stigmas become receptive, the flowers are known as 'protandrous' and if stigmas become receptive before the stamens produce viable pollens, it is known as 'protogynous'. Dichogamy has been reported in hermaphrodite (avocado, mango, ber and annona), monoecious (cashew, pecan nut, walnut, chestnut) and dioecious (persimmon, fig and pistachio) species. In walnut, occurrence of both protandrous and protogynous cultivars prevents self-pollination, which warrants cross-pollination. In mango, the duration of stigma receptivity is only 2 to 4 h, following anthesis.
4. **Stigmatic receptivity:** It is the ability of the stigma to support pollen germination and it limits the effective pollination period (it is defined as the number of days during which the pollination is effective in producing a fruit and is determined by the longevity of ovules minus the time lag between pollination and fertilization) in fruit crops. The stigmatic receptivity could be an important factor limiting pear flower receptivity (Sanzol *et al.* 2003).

5. **Abortive flowers or aborted pistils or ovules:** This occurs in the developing flowers, pistils and stigmas. Interference either in the development of the flower or in the full development of sex elements and their function may lead to unfruitfulness. Floral abortion is more common in indeterminate inflorescence as compared to determinate inflorescence. Pistil degeneration leads to unfruitfulness in certain cultivars of plum and ornamental pomegranate, while in strawberry, pistil abortion is so late that unfruitfulness does not take place.
6. **Non-viable pollen:** It is due to non-functional pollen or the ovule. Non viability or impotence of pollen results in unfruitfulness. Unfruitfulness in the case of Muscadine grape is due to defective pollen. In grapes, sterile pollen results from degeneration processes in the generative nucleus or arrested development prior to mitosis in microspore nucleus. In general, late flowering in apricot genotypes showed lower pollen viability than early flowering genotypes (Ruiz and Egea 2008).
7. **Sterility:** Some species have genes that prevent development of the pollen or the ovule. Generally, sterility is due to failure to obtain normal development of pollen, embryo sac, embryo and endosperm. Morphological sterility is due to rudimentary pistil or abortion of sex organs ovule degeneration. If one of the sexes is inactivated, cross pollination has to take place. Pollen sterility is common in peach cv. J. H. Hale. The proportion of sterile ovules varies between 22% in apricot and 98% in avocado (Verma and Jindal, 1997).

8.2.1.2 Genetic Influences

1. **Unfruitfulness due to sterile hybrids:** Hybridity is associated with sterility as well as unfruitfulness. The degree of sterility increases with wider crossing. The popular tangelo is a hybrid produced by crossing a grapefruit (*Citrus paradisi*) with a tangerine (*Citrus reticulata*). They are seedless or they produce seeds only with nuclear embryos (Gardner 1952).
2. **Incompatibility:** Incompatibility is defined as failure of viable pollen to grow down the style of flower of the same cultivar (self-incompatibility) or of the different cultivars (cross-incompatibility). Many cross-pollinating species exhibit self-incompatibility, so that fertilization by their own pollen is prevented through physical or biochemical factors. There are different degrees of self-incompatibility, and many self-incompatible species will produce a few fruits even when self-pollinated. Horticulturists have coined the terms "self-fruitful" and "self-unfruitful" to describe cultivars that can set commercial crops or

cannot set commercial crops (respectively) when self-pollinated. Self-incompatibility is more common in fruit crops like apple, pear, sweet cherry, almond, avocado, fig, mango, citrus, olive, etc. than cross-incompatibility (apple, pear, sweet cheery, European plums and almond). In fruit production, incompatibility may create a platform to create variation leaving no scope for inbreeding. In mango, self-unfruitfulness is reported in cvs. Dashehari, Chausa and Langra (Ram *et al.* 1976). Most pear (*Pyrus communis* L.) cultivars are impaired to set fruit under self-pollination because self-fertilization is prevented by gametophytic self-incompatibility system (Sanzol 2007). In loquat cultivars improved golden yellow, pale yellow, golden yellow; the pollen tube penetrated the stylar canal up to 1/4 to 1/3rd of its length and did not go further below, even after 72 hours of pollination. As such, this suggests incompatibility in loquat (Singh and Rajput 1964).

8.2.1.3 Physiological Influences

1. **Slow pollen tube growth:** Slow growth of the pollen tube results in unfruitfulness. Differences have been found in the rate of growth in selfed and crossed apples, pears, cherries and certain citrus fruits. This may be considered one type of incompatibility due to chemotropic or hormone influences. Besides this, fertilization should take place within a short time failing which abscission will take place at the base of the style, ovary, pedicel or peduncle and fruit setting does not take place.
2. **Premature or delayed pollination:** Premature or delayed pollination leads to unfruitfulness. Tobacco flowers are very susceptible to injury from premature pollination. When mature pollen grains are applied to immature pistils they germinate, penetrate the style, enter the ovule and if the ovules are not ready for fertilization the flowers fall. However, in case of oranges premature pollination did not have any deleterious effect whereas some injury was noticed in tomato. Lower fruit setting due to premature pollination was noticed in persimmon, pear, plum and peach. Similarly, if pollination is delayed the flowers fall without fruit setting. Delay in pollination for one or two days did not affect fruit set. However, further delaying may result into polyembryonic seeds in some species.
3. **Nutritive condition of plant:** Nutritive condition of plant just before or just after the time of blossoming is an important factor determining the percentage of flowers carrying for setting and for maturity. It may affect the pollen viability or fertility of pistils.

   i. Effect on pollen viability: There was significant difference in germination percentage of pollen collected from old apple trees and from strong young trees of the same cultivar.

ii. Effect on defectiveness of pistils: Exhaustion of tree by over bearing, drought or poverty of soil leads to production of defective pistils. Over bearing weakens the fruit tree and in coming season production is adversely affected. Close correlation was reported between defective pistils and unfruitfulness in American plums.

In case of *Vitis vinifera* carbohydrate deficiency is the common cause of flower drop. Due to carbohydrate deficiency flower abortion and ultimately unfruitfulness also occurs in green house grown tomatoes.

4. **Fruit setting of flowers in different positions:** Fruits borne on terminal growth have more competition in many fruit crops and mature and set under normal nutritional conditions but percentage of set is small. This positional competition takes place between fruits and branch as well as between different fruits influencing fruitfulness. Strong and Weak Spurs - Nutritional condition of spurs has positive correlation with fruit setting in apple. Spurs on vigorous limbs with large leaves set more fruits than those borne on weak limbs. More flowers ultimately lead to more fruit set and more flowers are generally borne on strong limbs. Likewise, flowers borne singly set fruits and mature as fruit and majority of those borne in clusters drop down. In the process of fruitifications, the embryo is more important for development i.e. if nutritive condition is favourable, it accompanies the development of the seed coat and fruit wall, if not, only the latter portions are in high degree retardation in development. Under insufficient nutrient supply the number of seed forming ovules are diminished and under extreme nutrition deficiency both fruit wall and large number of ovules are diminished leading to enabling to form seed. In case of greenhouse cucumbers, nutritional deficiency leads to arrest of growth of growing fruits depending upon the position of the fruits and time of pollination. If a few of the cucumbers are harvested remaining fruits resume growth. In case of strawberries producing bisexual flowers may lead to produce pistillate flowers if nutritional deficiency was observed. However, nutritive condition has indirect influence on compatibility. Some fruits like plum and cherry, bear on both shoots and spurs and it is to be expected that slightly different nutritive conditions are obtained in the different tissues and a distinctly heavier June drop occurs in shoot borne fruits (Gardner 1952). In apple, old spurs are less likely to initiate flowers than the young ones (Jackson 2003).

8.2.2 External Factors Associated with Unfruitfulness

The environmental conditions govern the life cycle of a fruit plant. These conditions greatly influence the flowering and fruit setting in the fruit plant.

Factors like nutrition, pruning, water supply, rootstock used, temperature, insect pests and diseases affect the fruiting in fruit trees.

8.2.2.1 Climate

i. **Temperature:** This is one of the most important factors, which governs the flowering, fruiting and fruit development. Temperature requirements are species specific. Variation in day and night temperature or extreme fluctuations, or continuously hot temperature adversely affects the pollination and fruit set in most of the fruits. For example, cloudy weather at full-bloom stage affect the pollination and fruit set in soft pears and plum. Poor setting in Sapota fruit is due to drying of flowers. Temperature affects the bee activity in an orchard. Bees help in the pollination and fruit setting. Unfavourable temperature may cause failure of any flowering as in the case of apples in Connor due to lack of sufficient winter chilling. In tropics, plants flowering in summer may experience retarded pollen germination due to high temperatures and low humidity.

   a) High temperature: Above 32°C, desiccation of the stigmatic surface and more rapid deterioration of embryo sac occurs (Jindal *et al.* 1993). As temperature increases, ovule senescence becomes faster in plum. At 15°C, only one ovule per flower remains viable by 8 DAFB for Italian, whereas for 'brooks' cultivar, higher temperature results to a decrease in ovule longevity (Moreno *et al.* 1992).

   b) Low temperature: In plum, cherry, apple, pear, etc., the temperature of 4.4°C or lower, completely check the blooming, fertilization and fruit set (*Jindal et al.* 1993). Fruit set and production of fruit in peach cv. Granda were absent or very low in green house grown trees in orchards. Nava *et al.* (2009) found that a rise of the diurnal air temperature in the interior of the greenhouse promoted a significant reduction in the production of pollen grains.

ii. **Cloudy Weather:** Cloudy weather is more dangerous than hails. The humidity makes the conditions most favourable to spread fungal diseases. Powdery mildew in mango and Umran ber usually appears in cloudy weather.

iii. **Rainfall:** Rains at the time of blooming period causes unfruitfulness by washing pollen grains, inhibit pollinators and cause spread of diseases and pests. Every fruit species needs a specific time period without rains at the time of blooming for successful pollinations. Late rains may prolong the vegetative growth and delay or reduce flowering in mango. Heavy rains may restrict pollinator activity, wash away pollen and prevent pollen germination. It directly affects fruit setting by disturbing the process of pollination and germination of pollen grains and staminal fertilization. In almond, washing treatment decreased the number of germinated pollen grains on the stigma mainly when the flowers were immersed before pollination and it seems to affect adhesion in forthcoming pollinations (Ortega *et al.* 2007).

iv. **Humidity:** Low atmospheric humidity causes drying of stigmatic secretions. Wet and humid weather favours anthracnose and poor fruit set in mango. The poor germination of pollen in almonds is attributed to damp weather during fruit set.

v. **Winds:** There are many fruits which get pollinated through wind (anemophily). For these plants, movement of air at the time of flowering is necessary to effect pollination. Most of the fruit plants are insect-pollinated (entomophilous). In such plants, wind hinders rather than helping in pollination. Hot wind (too) during May-June in North India desiccate the flowers or newly set fruits. Winds along with rain is more harmful. Wind also affects the fruitfulness indirectly by its effect on the pollinating agents, i.e., it controls pollination either by promoting wind pollination or by checking insect pollination. It is desirable in wind-pollinated fruit trees like hazelnut and walnut, but if its speed is too high, it is harmful because it results in low fruit set on exposed sides and heavy crop on other sides. Excessively speedy winds cause ovary abortion (Gardner 1952) and also make the stigma dry.

vi. **Frost/Freeze:** It is the most important factor for deciding the fruitfulness in an orchard. Frost injury can convert a regular bearing cultivar into an irregular bearing. Even orchards may not produce any fruit for two to three years continuously due to killing of branches caused by severe frost.

vii. **Hail Storm:** Hail storm has been found to be very harmful in hilly areas. Most of the apple crop was damaged by hails at fruit set. Hails kill flower buds and blossoms. There are areas which are prone to hails and freezes every year. Some areas are almost free from these hazards in the same zone.

viii. **Intensity of Light:** Light also plays a major role in the fruitfulness of an orchard. Strawberry plants develop pistils only when these are exposed to specific light intensity. In overgrown litchi and mango plantations due to overlapping shade the fruit set is reduced. Closely planted kinnow 10' by 10' or 10' by 20' at full growth stage bears poorly due to poor light penetration. Even fruits do not develop proper colour at maturity. Reduced illumination due to close planting, overcrowding of branches or shade will often reduce flowering. When long day plants of northern latitudes do not flower owing to the absence of the critical length of day, they can be made to flower by providing artificial light. On an orchard scale such treatments are not light that affects fruitfulness indirectly by its effect on photosynthesis. Light is a pre-requisite for photosynthesis and low light intensity or its duration reduces the carbohydrates reserves in the trees. In addition to this, poor light conditions promote fruit abscission.

8.2.2.2 Disturbed Moisture in Soil

Soil moisture is one of the key factors to production of fruits. The excess soil moisture as well as low soil moisture conditions at the time of flowering and fruit set encourages abscission layer formation, leading flower and fruit drop.

8.2.2.3 Nutrition

When balanced nutrition is not given, the plant's growth and development are affected. The manures and fertilizers need to be applied one to two months before flowering and in split doses after fruit setting. Imbalance in the nutrients certainly causes unfruitfulness or flower drop. Heavy nitrogenous manuring at the time of flower bud initiation often reduces flowering by promoting vegetative differentiation. The practice should be given up. Over bearing in the previous season exhausts the tree and reduces subsequent flowering as in mango and most biennial bearing trees. Lack of nutrition as in weak shoots causes fall of flowers before and after fruit set. A spray of urea after fruit set will help the development of fruits. Lack of sufficient reserves of carbohydrates in shoots may cause sparse flowering and poor set (shoot bunches of grapes). Adverse growth features like water suckers will result in a drain on the tree and reduce flowering not only on themselves but also on other branches of the tree. Such shoots arise when big branches are pruned. Then it is necessary to cut big limbs, they should always be set to a strong lateral but not stimulate a dormant bud. Late irrigation following a long drought may cause the production of water shoots. Deficiencies of elements are sure cause of reduced flowering as well as set. Heavy manuring and severe pruning during the pre-bearing period will prolong flowering as well as set.

What is C/N ratio and its relationship with fruitfulness?

The balance between carbohydrate and nitrogen in plant is known as C/N ratio.

- The fruit bud formation, setting of fruit, and its development mainly depends on C/N ratio in the branches of the trees at certain critical stages of seasons.
- Plants with adequate nitrogen and high carbohydrate accumulation are found to make moderate growth and produce satisfactory crop.
- Plant with excess of nitrogen and moderate carbohydrates are found to grow vegetatively at the expense of fruit production.
- The accumulation of carbohydrates in the branches may be either due to their manufacture or due to less rapid utilization. In plants, which are weak in their growth, carbohydrates may be made to accumulate by supplying them more nutrients and water.

Certain horticultural practices like branch bending, ringing, scoring etc. increase C/N ratio in branches.

On the basis of nutritional status, fruit trees have been divided into 4 general groups.

C/NNNN: Plants in this group make poor growth and bear little crop or no fruit. Shaded or over crowded trees falls in this category.

CC/NNN: Excess of nitrogen is present with sufficient amount of carbohydrates for strong and succulent growth.

CCC/NN: Excess of carbohydrates leads the plants to its favourable growth and fruit set. Under such condition optimum fruitfulness can be obtained.

CCCC/N: Excess amount of carbohydrates and small quantity of nitrogen leads to the poor growth and small amount fruits, neglected trees fall under such category.

8.2.2.4 Rootstocks

Rootstocks affect the scion cultivars physiologically. Quince rootstock induces dwarfing in pear due to formation of inverted bottleneck, whereas D-4 produces very vigorous pear plants. Trifoliate orange and its hybrids Troyer and Carrizo produced dwarf citrus plants than on Jatti Khatti and Kharna Khatta. The grafted plants produce early crops than those raised through seeds, this way fruiting is affected by the rootstocks. Use of inter-stock reduced the juvenile period by two years in 'Leconte' and other soft pears. Patharnakh (*Pyrus pyrifolia*) inter-stock was better than root suckers (*Pyrus calleryana*).

8.2.2.5 Pruning

The deciduous trees are judiciously pruned each year. Un-pruned grapes bear little crop with small bunches as compared to pruned vines. Un-pruned peaches bear little fruit and are prone to limb breakage. To improve fruiting, right pruning is given every year. Similarly, summer dormant ber is pruned in May-June to get more fruit. The intensity of pruning varies from cultivar to cultivar and species to species.

8.2.2.6 Plant Age

Some fruit plants have long juvenile period than others. These plants cannot be made to bear fruit early. The *Citrus medica* and *C. jambhiri* plants bear only male flowers in the first few years. Slowly hermaphrodite flowers appear with age. Young grape vines produce less pollen than the aged vines of the same cultivar. Very healthy plants bear less flowers than semi- healthy plants. Declining plants bear profusely than healthy trees.

8.2.2.7 Chilling Requirements

Some fruit trees need desired chilling hours requirement for spur formation and flowering. Bartlette and conference pears do not flower in plains; whereas low chilling requiring 'Leconte' and other pears bear heavy fruit. Apple remains vegetative for long in lower hills than higher altitudes. Southern cultivars of mango shoots get killed due to frost or low temperature in north, hence no fruiting. In 'Leconte' pear profuse flowering is observed every year but fruit set is irregular due to change in season.

8.2.2.8 Spraying at Full-bloom

Normally insecticide sprays at flowering time are not recommended. However, sometimes due to prolonged flowering particularly in litchi, mango and pear, it becomes necessary to control aphids, etc. Hence, insecticide is sprayed, which adversely affect the fruit set. Spray may be advanced or delayed for a week or so.

The use of pesticides can kill bees, therefore reducing pollination. Some pesticides can also be toxic to delicate flowers causing abortion and loss of fruit. Pesticides spray effects on receptive stigmatic surface showed varying degree of injury and range from minor surface wrinkling to degeneration of stigma papillae. Controlled pollination for 1 h, after pesticides sprays, results in an inhibition of pollen germination and tube growth (Wetzstein 1990). Commercial fungicides containing captan, dinocarp, sulphur and triforine sprayed on undehisced anthers of several apple cultivars reduced the viability of pollen, impaired pollen release, killed pollen when sprayed onto dehisced anthers (Ruth *et al.* 1978). Pollen germination and tube growth were drastically suppressed by the spray and almost no fruit set was observed on the treated inflorescence and there was a highly significant difference between fruit set on exposed and protected cluster (Legge and Williams 1975).

8.2.2.9 Biological Causes

Absence of pollinating agents can be a reason for low fruit set in several fruits. Rearing bee colonies in orchards, besides being a subsidiary source of income greatly helps fruit set.

When specific insects' symbiotic adaptations (like the blastophaga for fig) concerned with the pollination, they must be reared (by growing Capri fig trees in this case). A wooden needle seems to perform the duties of the blastophaga quite as efficiently in promoting set of fig fruit.

Pests like the mango hopper which directly attacks the flowers obviously reduce the fruit set. Others which feed on leaves reduce the photosynthetic surface impair production of carbohydrates and thus reduce flowering. Several fungal diseases do the same thing; suitable remedial measures should be taken to protect the trees.

8.2.2.10 Cultural Causes

The commonest cause of poor flowering in house gardens is excessive irrigation which restricts aeration of roots and causes sickly symptoms. Weeds and intercrops may compete with the main crop for nutrition and water in low rainfall areas. Ploughing or deep cultivation at flowering time will result in drop of flowers and should be avoided. Severe pruning of large limbs which encourages production of water shoots should be avoided.

8.3 Remedies to Control Unfruitfulness

8.3.1 Pruning and Thinning

Problem due to uneven distribution of flowers on tree should be managed through thinning and crop regulation. The pruning should be with regard to bearing habit of the fruit tree; the pruning should be up to some fruitful buds (in grapes). Harder or light pruning will reduce fruiting. A balance of vegetative and fruiting wood should be maintained in plants bearing on past seasons wood. Pruning should not be delayed till the new growth is resumed. In crops requiring regular pruning standard practices will have to be adopted based on crop, cultivar and its phenology. Heavy manuring and severe pruning during the pre-bearing period will prolong flowering as well as set. Pruning should be done while branches are young, preferably by rubbing of axillary buds themselves by frequent observation of the plants. Problem due to overbearing can be managed through thinning at appropriate stage. Severe pruning of large limbs which encourages production of water shoots should be avoided. If it is absolutely essential, the branches may be cut to a strong lateral. Wrong pruning techniques may also cause reduction of flowering.

The main techniques for controlling the vigour of fruit trees and increasing their relative fruitfulness are the use of dwarfing rootstocks, compact short-internode scions and training and pruning systems of trees which give horizontal or wide-angled branches. As such, growth retardants are also used. The influence of pruning varies with the amount, season and kind of pruning. Judicious and proper pruning is needed to improve the fruit set and fruit retention on the trees. Severe pruning promotes too much vegetative growth and hence reduces the productivity. Experiments revealed that pear on *P. communis* and 'E.M. Quince C' rootstocks showed early fruit thinning to 1 fruit per cluster and increased the ultimate percent fruit set of 'Comic' girdling and cluster thinning in 'Anjou' set beyond either treatment (Westwood 1974). In mango, the highest number of new flushes per shoot was achieved with severe pruning and spraying of GA_3 at 100 ppm (Sheban 2009). Singh and Daulta (1986) found that pruning up to 12- bud improved fruit set significantly and it is suggested that 'Sharbati' cultivar responds well to light pruning.

8.3.2 Use of Pollinizers

Some cultivars are sterile, self-unfruitful cross sterile. In such cases block of other cultivars which provide compatible pollens should be established. Polliniser trees may be interplanted in the orchard/in suitable numbers in different pockets. Ten percent of male plants of papaya is required for hood fruit set. In case of problems of pollination due to heterostyly, dichogamy, incompatibility, sterility, embryo abortion, hybridity, etc., a mixture of cultivars should be grown by introduction effective pollinizer cultivars and pollinators (Honey bees). Pollen transfer may present an application problem in fruits which are self-incompatible. Many apple cultivars are at least partially self-fertile, especially under warm weather, but in most cases, fruit-sets are improved by the use of pollinator cultivars and bees. Cherries and plums also need cross-pollination and should be interplanted with pollinizer cultivars. Breeding and selection of self-fertile cultivars or clones of cherry and apple may reduce difficulty or achieve satisfactory pollination, especially in cool marginal areas of fruit production where temperatures at blossoming time are sub-optimal both for bee activity and for pollen tube growth. Until such improved cultivars become available, provision of suitable pollinizer cultivars and bee hives can be of great help in ensuring satisfactory fruit-set. There is an increasing tendency for using crab-apples, as pollinisers of these occupy very little space in orchards. It is advisable to have a number of pollinizer cultivars with widespread flowering dates to ensure cross-pollination. One of the basic requirements for setting fruit is an adequate requirement of compatible pollen. With most tree fruit crops, the need for cross-pollination is recognized. Pollinating insects are necessary for fruit set on all cultivars, and most cultivars will benefit from cross-pollination. Under general conditions, the closer a tree is to a pollinizer, the better fruit set will be. Cross-pollination necessitates the availability of sufficient quantity of compatible pollen, as pollinizer cultivars flower synchronously with the main cultivar and suitable agent for the successful and effective transfer of pollen.

8.3.3 Introduction of Pollinators

The population of natural pollinators has gone down due to indiscriminate use of pesticides and deterioration of the ecosystem. The bee-managed pollination is very limited and the available bee hives, during bloom, hardly meet 2 to 3% of the demand. In red delicious apple, it was found that sequentially increasing the density of colonies increases the amount of cross-pollination and a high proportion of top workers. However, the increased pollination efficiency results in high fruit set and higher yield (50 to 100%) in treated plots (Stern *et al.* 2001). Rearing of honeybee colony will be effective for transfer of pollens.

8.3.4 Proper Nutrition

Maintenance of critical nutrient status in tree leaves for optimum crop production by adopting correct nutritional programme is based on plant and soil analysis. Overbearing in the previous season exhausts the tree and reduces subsequent flowering as in mango and most biennial bearing trees. A complete manure mixture applied at the growth flush following the harvest will be helpful (June manuring in mango). Balanced supply of nutrients is always desirable for realizing optimum fruit production. Generally, it is advocated that for application of fertilizers, a few days before emergence of blossoms is generally believed to favour flowering and fruit set. Nitrogen application after terminal bud formation led to the development of flower with enhanced embryo sac longevity (Jackson 2003). Pollen tube growth in pear was significantly stimulated by increasing concentration (25 – 200 mg/l) of boric acid and the values are significantly higher at 200 mg/l concentration of boric acid (Lee *et al.* 2009). In walnut cv. Local selection, the highest fruit set (32.50), fruit retention (41.35) and nut yield (4.02 kg/tree) were recorded under foliar application of H_3BO_3 + Paras (0.1 % + 0.6 ml/L) (Tomar and Singh 2007). However, excess use of manures and fertilizers may produce abnormal flowers.

8.3.5 Application of Plant Growth Regulators

By application of plant growth regulator growth rate of pollen tube into style can be enhanced. Unfruitfulness due to slow growth of pollen tube, premature and delayed pollination, use of plant regulators can be effected after standardization in terms of chemical concentration and timing of application. The unfruitful behaviour of several fruit plants can be overcome by the use of plant growth regulators which may be due to decreased fruit set and abscission at various developmental stages.

8.3.6 Use of Suitable Rootstocks

There can be as much as 50% or more difference in the yield of a given cultivar grown on different rootstocks. The reasons for such an effect can be traced to difference in tolerance to adverse soils, in resistance to pests or in uptake of nutrients. Four rootstocks namely: M9, M7, M4 and M1 induced 50% or more bloom in the fifth year in Starking delicious apple and resulted in higher yield efficiency by controlling tree size. (Wani *et al.* 2010).

8.3.7 Irrigation Management

Irrigation management would be key role in situations with drought and waterlogged conditions. Excessive moisture supply, deficient moisture supply or fluctuating moisture at the critical stage of flowering and fruit set may develop into failure one, hence adequate and timely irrigation frequencies be maintained to regulate moisture supply. The commonest cause of poor flowering in house gardens is excessive irrigation which restricts aeration of roots and causes sickly symptoms. Increase of intervals of irrigation and provision of drainage are the remedies. Growing cover crops or mulching the basins can overcome the low soil moisture. Flooding of the orchards at flowering time should be avoided. The first irrigation after a drought should always be sparing later ones being more liberal.

8.3.8 Other Remedial Measures

It is important to analyse the problem and then corrective measures should be suggested. Basically, the planning should be so done that future is problem free and then should be followed by adoption of correct package of practices.

- Choice of the crop and cultivar should be made on the basis of climatic and edaphic conditions of the site of orcharding.
- Before planting an orchard, soil should be brought to optimity by incorporating organic matter, amendments and nutrients based on soil analysis.

- Ploughing or deep cultivation at flowering time will result in drop of flowers and should be avoided.
- The problem due to old age could be overcome by replanting or rejuvenation of old trees.
- Unfruitfulness due to insect and pathogens should be managed through effective plant protection measures following integrated approach.
- Problem of unfruitfulness due to tendency of alternate bearing should be overcome through replacement of regular bearing cultivars and crop regulation.
- By maintaining the proper nutritive health of the plant and attending CCC/NN ratio, it will produce optimum fruitfulness.
- Special horticultural practices like bahar treatments, ringing may improve fruit set by bringing the plant in CCC / NN ratio.
- Wind breaks should be planted on south-west side of orchard to protect the plant from high wind velocity with high temperature and low humidity.
- Root pruning and restricted irrigation may be helpful in reducing vegetative vigour and inducing formation of male flowers.
- Lack of sufficient reserves of carbohydrates in shoots may cause sparse flowering and poor set (shoot bunches of grapes). Ringing and girdling may help. But it should not be continued as a regular orchard practice.
- A composite mineral spray at flush time will usually be very helpful. If deficiency is due to alkalinity of the soil, suitable reclamation measures should be adopted.
- Weeds and intercrops may compete with the main crop for nutrition and water in low rainfall areas. Removal of weeds and adequate manuring to meet the demands of both the fruit crop as well as the intercrops are helpful.
- Intercrops which clash with the irrigational and manurial requirements of fruit crops both in respect of quality and time of application should be avoided.

References

Baldwin EA (1993). Citrus fruit. In: Seymour GB, Taylor JE, Tucker GA (eds), *Biochemistry of Fruit Ripening*, pp.107-149. Chapman & Hall, London.

Erickson LC and Brannaman BL (1960). Abscission of reproductive structures and leaves of orange trees. *Proc. Am. Soc. Hort. Sci.*, **75**: 222-229.

Davis FS and Albrigo LG (1994). Citrus. In: *Crop Production Science in Horticulture Series*, pp.134-135. CAB International, Wallington.

Gardner VR, Bradford FC and Hooker HD (1952). *The Fundamentals of Fruit Production.* McGraw-Hill Book Company, INC., New York, Toronto, London, pp. 536-554, 643-685.

Ghawade SM, Panchbhai DM, Jadhao BJ, Katole SR and Dadmal SM (2002). Flowering behaviour of Kagzi lime (*Citrus aurantifolia* Swingle) under sub-humid tropical climate. *Agric. Sci. Digest.*, **22** (4): 252-254.

Ghosh P (1998). Fruit weather of India. In: RK Arora and Ramanatha Rao (eds). *Tropical Fruits in Asia.* IPGR., 3-5 pp.

Goldschmidt EE and Huberman M (1974). The coordination of organ growth in developing Citrus flowers: a possibility for sink type regulation. *J. Exp. Bot.*, **25**: 534- 541.

Goldschmidt EE and Monselise SP (1977). Physiological assumptions toward the development of a citrus fruiting model. *Proc. Int. Soc. Citrus,* **2**: 668-672.

Jackson JK (2003). *Biology of Apples and Pears.* Cambridge University press, UK pp. 268-340.

Lee SH, Kim WS and Han TH (2009). Effect of post-harvest foliar boron and calcium applications on subsequent season's pollen germination and pollen tube growth of pear (*Pyrus pyrifolia*). *Scientia. Hort.*, **122**: 77- 82.

Legge AP and Williams RR (1975). Adverse effect of fungicidal sprays on the pollination of apple flowers. *J. Hort. Sci.*, **50**: 275-277.

Monselise SP (1986). Citrus. In: Monetise SP (ed), *Handbook of Fruit Set and Development*, pp.87-108. CRC Press, Boca Raton.

Naik A (1949). *South Indian Fruits and Their Culture.* P. Varandachari & co., Madras. 94 pp.

Ortega E, Dicentia F and Egea J (2007). Rain effect on pollen stigma adhesion and fertilization in almond. *Scientia. Hort.*, **122**: 345-348.

Ram S, Bist LD, Lakhanpal SC and Jamwal IS (1976). Search of suitable pollinizers for mango cultivars. *Acta. Hort.*, **57**: 253-264.

Ruiz D and Egea J (2008). Analysis of the viability and correlation of floral biology factors affecting fruit set in apricot in a Mediterranean climate. *Sci. Hort.*, **115**: 154-163.

Sanzol J, Rallo P and Herrero M (2003). Stigmatic receptivity limits the effective pollination period in 'Agua de Aranjuez' pear. *J. Amer. Soc. Hort. Sci.*, **128** (4): 458-462.

Sanzol J (2007). Self-incompatibility and self-fruitfulness in pear cv. 'Agua de Aranjuez'. *J. Amer. Soc. Hort. Sci.*, **132** (2): 166-171.

Sharma S, Sehrawat SK and Sharma KD (2017). Studies on time and duration of flowering, floral bud development and morphology of guava (*Psidium guajava* L.) under semi-arid region of India. *Int. J. Curr. Microbiol. App. Sci.*, **6** (12): 4176-4186.

Sheban AEA (2009). Effect of summer pruning and GA3 spraying on inducing flowering and fruiting of zebda mango trees. *World. J. Agri. Sci.*, **5** (3): 337-344.

Singh D and Daulta BS (1986). Effect of pruning severity on growth, flowering, fruit set, fruit drop, the extent of sprouting and flowering at various nodal positions in Peach (*Prunus persica* Batsch) cv 'Sharbati'. *Haryana. J. Hort. Sci.,* **15** (3-4): 200-205.

Singh JP and Rajput CBS (1964). Pollination and fruit set studies in loquat (Eriobotrya japonica). *Indian J. Hort.*, **21** (2):143-147.

Singh LB (1971). *The Mango: Botany, cultivation and utilization.* Leonard Hill Ltd., London.

Singh NP, Dhillon WS, Sharma KK and Dhatt AS (2006). Effect of mechanical deblossoming on flowering and fruit set in pomegranate cvs. 'Ganesh-1' and 'Kandhari'. *Indian. J. Hort.* **63**(4): 383-385.

Singh SN and Tomar BS (1949). *Indian Farming*, **10**: 532-38.

Stern R, Goldway M, Zisovich AH, Shafir S and Dag A (2004). Sequential introduction of honeybee colonies increases cross pollination, fruit set and yield of Spadona Pear (*Pyrus communis* L.). *J. Hort. Sci. Biot.*, **79** (4): 652-58.

Suranyi D (1976). Differentiation of self-fertility and self-sterility in *Prunus* by stamen number/ pistil length ratio. *Hort. Sci.*, **11** (4): 406- 407.

Thakre M and Nayan Deepak G (2015). Bearing Habit of Fruit Crops. In: *Fundamental of Fruit Production*, Division of Fruits and Horticultural Technology, Indian Agricultural Research Institute, New Delhi, edited by Usha K, Thakre M and Goswami AK and Nayan Deepak G.

Tomar CS and Singh N (2007). Effect of foliar application of nutrients and bio-fertilizers on growth, fruit set, yield and nut quality of walnut. *Ind. J. Hort.*, **64** (30): 271-273.

Ullah MA and Haque MA (2008). Studies on fruiting, bearing habit and fruit growth of jackfruit germplasm. *Bangladesh J. Agril. Res.*, **33** (3): 391-397.

Vakis RM and Williams RR (1978). Fungicide toxicity to apple pollen in the anther. *J. Hort. Sci.*, **53**: 91-94.

e-resources

e-Learning portal on Agricultural education, https ://ecourses.icar.gov.in

https://www.yourarticlelibrary.com> crops

https://researchgate.net

Outcomes Assessment

Part A

Answer the following questions true or false.

1. Unfruitfulness can be due to lack of balance between vegetative growth and fruiting. True/False
2. Absence of pollinating agents cannot be a reason for low fruit set in several fruit crops. True/False
3. The commonest cause of poor flowering in house gardens is deficit irrigation. True/False
4. Rearing of honeybee colonies is effective for transfer of pollens. True/ False
5. Light does not play any role in the fruitfulness of an orchard. True/False

Part B

Answer the following questions.

1. Define fruitfulness.
2. What is chilling requirement?
3. There can be as much as — or — difference in the yield of a given cultivar grown on different rootstocks.
4. Most of the fruit plants are insect/wind-pollinated.
5. In mango, self-unfruitfulness is reported in cultivars —, — and —.

Part C

Write an essay on each of the following.

1. Remedies to control unfruitfulness.
2. External factors associated with unfruitfulness.
3. Internal factors associated with unfruitfulness.

9

Rejuvenation of Old Orchards

9.1 Introduction

India is a vast country and is gifted with a variety of soils and climates. Thus, almost all kinds of fruits can be grown successfully in this country. It is a matter of great surprise that in spite of provision of adequate resources, the per capita consumption of fruits here is perhaps one of the lowest in the world.

The existing orchards are not able to meet the present requirements of the country. Poor selection of planting material, haphazard plantation and poor management has made many orchards uneconomic. The term 'Rejuvenation' means renewal or making new or young again.

As applied to the orchard tree it would mean restoring the productive capacity of the fruit trees. The meaning of 'Rejuvenation' according to Chamber's dictionary is 'to recover youth character or to grow again'. Obviously, this would apply to those plants which have attained a stage where they are no more profitable from the grower's point of view.

9.2 Causes Which Make the Orchard Uneconomic

Survey of uneconomic orchards would reveal the following defects:

i. Growth of wild shrubs and grasses

Wild shrubs and grasses continue to grow without any check under the tree and between the rows. At a stage they develop into bushes here and there in the entire plantation. It has been realized that vegetation removes quite a larger quantity of moisture from the soil and also ties up nitrogen so that it is not available to the trees at the time, though it may be returned later when the weeds die and decompose in the orchard soil. Weeds harbour insects, pests and diseases too. When they become large enough, they interfere in the orchard operations and some of them climb up the tree and by shading the leaves or due to their parasitic nature cause great damage to the main fruit crop.

ii. Overcrowding and unsystematic planting

In some of the orchards no regular or systematic planting is followed. The trees are also found to be planted so close that they form a dense overhead canopy which does not allow sufficient light to reach at different parts of the fruit trees uniformly. The lower branches remain under perpetual shade. As a result of it the bigger branches remain devoid of small twigs or branchlets and ultimately this abnormality reduces the fruiting surface. In such cases, trees for want of space grow tall and bear fruits almost on periphery only. These fruits in hotter localities are affected adversely by sun burn. This way the growers unfortunately suffer a double loss because on one hand the quality of the fruit and on the other the total produce, both are considerably reduced. Free circulation of air is also checked and this results in development of pests and diseases.

iii. Inferior cultivars

The poor selection of planting material at the time of planting will cause a great loss throughout the life of the orchard. In most of the orchards, trees have been raised from seeds and of inferior cultivars. Seedlings of poor cultivars are potentially poor and their performance is subjected to a wide range of variation so that even if they grow and function normally, they do not produce a crop that can fetch good price in the market.

iv. Damage due to adverse weather conditions, rodents and other enemies

Trees with damaged trunks, broken limbs and big hollow in the trunks are not uncommon in such orchards. Some trees may be found standing with only half their base. Many trees live for hundred years or more and grow to huge dimensions. Old trees with the advanced age have a tendency to develop hollows in their trunks and main branches start from a crotch or from the point of bifurcation. The branches which cross one another rub together when fast wind blow and injure their bark. Rain water is deposited in the hollows produced and cause the bark and wood inside to rot. As the process continues year after year, larger cavities are formed. In course of time the main trunk develops a hollow in which three to five persons could enter at a time.

v. Infestation of pests, diseases and parasites

The orchards which have been established in neglected localities are seriously attacked by number of pests, diseases and plant parasites. These enemies are so rampant that even in orchards otherwise in good condition, the production is very poor. These organisms devitalize the plants and destroy flowers and fruits in all stages of development.

9.3 Practices For Rejuvenation

The first step while going to rejuvenate an uneconomic orchard should consist of a thorough examination of the causes which led to the tree in poor state. Efforts should then be made to rectify the defects. An attempt has been made in the following paragraphs indicating various suitable suggestions helping for rejuvenating an uneconomic orchard. Practices to be followed in advance before the trees reach at need of rejuvenation to protect the orchard from this malady have also been explained side by side.

i. Soil management practices

It is very important that the soil should intelligently managed so that it may be maintained in a condition suitable for meeting the needs of the trees without undue expenses. There should be proper vigilance regarding the physical condition of the soil and also about moisture and nutrient contents. These depend largely on the practices of cultivation, irrigation, drainage, manuring etc. Conservation practices should also be followed to maintain the fertility and topography of the soil.

ii. Cultivation

It involves the elimination of wild shrubs and grasses from the land left vacant in between the trees. This practice has been found to prevent the surface soil from becoming hard. Hard soil greatly adds to the difficulty of irrigation and interferes with the growth of roots of main trees. Manures, fertilizers, green manure and other organic matter should be mixed with the soil by means of regular cultivation. It would facilitate in aerating the soils and would kill a number of insect pests and other harmful organisms present in the soil.

iii. Irrigation

The water relation of the plant is of extreme importance both for vegetative growth and for fruit production. Irregular irrigations may cause fruit drop, fruit splitting and other disorders. Application of manure should be followed by watering so that the nutrients are readily absorbed. Soil around the tree base should be raised to form a platform so that water does not touch the bark of the trunk directly. In areas of scanty rainfall or limited water supply, special methods of moisture conservation such as mulching may be practiced.

iv. Drainage

Drainage is as important as water supply. The absorption of nutrients is affected to a great extent by the aeration of the soil which is affected by drainage. After prolonged stagnation of water roots below the soil surface die and they may not be replaced by new ones. In poorly drained orchards, effective steps like provision of drains etc. should be taken, so that there is not any interruption in the physiological processes of the plants.

v. Manuring

Fruit trees live long, standing at the same place and bear crops year after year. In doing so, they constantly remove plant food from the soil unless an adequate return is made to the soil to replenish the plant part removed. Cultivation of green manure crops during early rains and turning them in the orchard soil and creating proper provision for their decomposition would greatly help in increasing the organic matter of the soil, and this will directly influence the supply of various nutrients. If green manuring is not possible, a heavy dose of FYM or compost may be spread throughout the orchard. Besides, individual trees should be manured with quick acting fertilizers. When the tree fails to respond to this treatment, a dressing of super phosphate and sulphate of potash or other similar mineral fertilizers may be done. Soil lacking in minor elements may also be corrected by adding certain chemicals possessing the same elements. Supply of trace elements is found to be more economic and effective through foliar sprays. This would enable to check over- fruiting, fruit drop, fruit splitting, diseases and many other disorders.

vi. Thinning of overcrowded trees and filling of gaps

Overcrowded trees should be thinned to proper distance. Though it might create a loss of some trees, but would also greatly benefit those plants that are left behind. However, it is not useful to see many trees missing. In those cases, gaps should be filled up by trees of superior cultivars. While undertaking these operations attention should be paid to bring the trees in lines according to system of planting originally kept in view. Sometimes certain cultivars of citrus which have been propagated vegetatively for many generations become somewhat lacking in vigour.

vii. Pruning technology for rejuvenation and restoration of productivity

The disbalanced root-shoot ratio can be corrected by judicious pruning. The branches which have died or broken or one which interferes with natural growth, and water sprouts should be removed. If the exposed wounds are big enough, they should be disinfected with arsenical paints. Old bearing trees that have reached their middle age have become somewhat low in vigour due to constant cropping or neglect, should be pruned heavily. Such trees respond better to a heavier pruning because of their reduced vigour. This treatment is to be supplemented with a heavy dose of manure later on. Rejuvenation in guava and peach plantation is achieved by heading the trees back almost to the base of the trunk. The low stems thus left produce vigorous growth and fair head is attained in two years. Grape vines are also similarly rejuvenated when they loss vigour. The deciduous plants respond better to severe pruning whereas evergreen ones are said to grow slowly for some time afterwards. Root pruning also sometimes restore the vigour of unproductive plants.

Mango having a long gestation period of orchard establishment (nearly 10 years), replacement strategy will have serious impact on the socio-economic conditions of growers whose livelihood support system is mainly dependent on mango production. Also, after uprooting instead of replacement with new orchard, opting for new cropping pattern and alternate land utilization pattern will have adverse impact on environment as well as the acreage under mango cultivation.

Keeping in view the above constraints, research endeavour at CISH, Lucknow focussed upon providing new productive life to existing old and unproductive orchards through rejuvenation technology. At CISH, experimentation on pruning of branches gave encouraging results in restoring the productivity potential. Various severities of pruning (pruning at different orders of branching) and time period of pruning were tried to standardise the most appropriate height and time period of pruning. Technique for rejuvenation of old and unproductive orchards of mango through pruning and intensive after care management was standardised at the Institute after concerted research efforts for nearly a decade. The technique aims at pruning of undesired branches for inducing development of umbrella like open canopy of healthy shoots. Open canopy ensures better light penetration and interception and improves flowering and fruiting potential of shoots. Pruned trees attain canopy of healthy shoots in two years and from third year onwards they start bearing fruits.

According to the standardised technique, branches are pruned from a height of 4-5 metres from ground as per the architecture of individual tress, during December. Before undertaking pruning work, umbrella like frame work of individual tress is planned. About four to five main branches with outward growth are considered for constituting the basic frame work of the tree for development of canopy. These four to five selected branches are marked with white chalk by making a ring around the branches at a distance of 75 cm from their base. Other criss-crossed, intermingling, overcrowding, dried and diseased branches are marked with red chalk for complete removal. Marking of branches in the orchards to be pruned for rejuvenation may be completed by November. During December, firstly the red marked branches should be completely removed by pruning from their base. Later, branches marked with white chalk should be pruned at a distance of about 75 cm from their base. Development of canopy is promoted on these branches.

Pruning can be done either with manual saw or power-operated saw (electric or petrol operated saw). Power operated saw has been found to be effective, labour and time efficient and economical for pruning work. In manual saw process of pruning, often problem of bark splitting or debarking at cut end of the branch arises due to weight of falling pruned branch. As a result, tree is exposed to microbial infection. Hence, to avoid it, pruning should be initiated firstly from the lower surface of the branch. Up to 15-20 cm deep cut should be made and later pruning may be done from the upper surface of the branch. Tall and heavy branches should be tied with rope and safely pulled for guiding their smooth falling on ground as well as to avoid splitting of branch.

Immediately after pruning, fungicidal paste prepared by mixing 1 kg copper oxychloride, 250 g castor oil and water as per requirement, should be applied on the cut surfaces to check microbial infection. Application of a cow dung paste or biodynamic tree paste, prepared by mixing cow dung, orchard soil, same in equal proportion with cow horn manure (B.D. 500) has been found to be quite effective in checking microbial infection

viii. Alternate row pruning

Pruned trees come into flowering and fruiting after 2 years of pruning and so growers have to suffer economic loss by missing crops for two years. Hence, technology was refined to undertake pruning work in alternate rows of tree in the orchard. Technology assessment and refinement trials in growers' orchards with alternate row pruning system gave encouraging results. As a result of alternate row pruning, availability of light to unpruned trees in two adjacent rows is greatly improved and their fruiting potential is increased by 2-3 times. Enhanced production from unpruned trees compensates the loss of produce from pruned trees. Economic loss is further compensated by income from sale of wood from pruned trees and additional return from intercropping in space opened up due to pruning. Once the pruned trees in alternate row attain canopy and come into fruiting, the remaining trees are subjected to pruning. Thus, alternate row pruning cycle minimizes the risk, compensates economic loss and makes the technology more compatible and adoptable among the growers, particularly, those with small size of holding.

ix. Integrated orchard management

Pruned trees need intensive care of nutrition and irrigation besides cultural practices like hoeing, weeding, etc., for survival, emergence of new shoots and development of ideal canopy. Application of 2.5 kg Muriate of Potash besides 120 kg well decomposed Farm Yard Manure (FYM) is recommended for each pruned tree. Half dose of urea with full dose of Single Super Phosphate (SSP) and Muriate of Potash is applied during the end of February. The remaining half dose of urea is applied during the end of June. Full dose of FYM (120 kg/tree) should be applied in the first week of July. Manures and fertilizers should be applied in the basins prepared around the trees. Basins and irrigation channels should be prepared during January after completion of pruning work. Hoeing and weeding in the basins must be done before application of manures and fertilizers.

After emergence of new shoots on pruned trees, optimum soil moisture should be maintained otherwise, new shoots may dry up. Therefore, depending upon temperature and soil moisture status, pruned trees must be irrigated at an interval of 15-20 days from March till the onset of monsoon. Mulching may be followed for conserving soil moisture in the basins during April to June. Dried grass, mango leaves, straw or black polythene sheet may be used for mulching. All the management practices as mentioned above, must also be followed for unpruned trees in alternate row to secure higher yield from them.

x. Inter Cropping

Pruning of trees in alternate rows make the orchard space open with greater availability of sunlight. Interspace between the rows of trees can be successfully utilized for intercropping. Flowers (marigold, gladiolus, etc), vegetables (cucurbits, okra, lobia etc, during *kharif* season and potato, pea, brinjal, cauliflower, cabbage etc, during *rabi* season) and spices (ginger and turmeric) are ideal intercrops. Intercropping generates additional employment and income for farm families.

xi. Thinning of newly emerging shoots

Three to four months after pruning, i.e., during March-April there is profuse emergence of shoots on pruned branches. If they are allowed to grow, there is undesirable competition among shoots for space, light, nutrition and growth. Consequently, dense and brushy canopy of unhealthy shoots with poor bearing potential develops on pruned trees. Therefore, selective and regular thinning of shoots is essential for facilitating development of open and spreading canopy of healthy shoots. Only outwardly growing 8-10 healthy shoots are retained per branch and the rest are removed so that they get proper nourishment and develop into ideal canopy.

Thinning operations are undertaken during June and August. Copper oxychloride fungicide (3 g per litre water) should be sprayed immediately after thinning. It helps in managing the incidence of anthracnose disease. Foliar spray of 2 per cent urea during October is beneficial for healthy growth of shoots.

xii. Integrated Pest Management_

Intensive care for prudent management of insect pests and diseases is highly essential for ensuring survival of pruned trees as well as health growth of shoots. Microbial infection in cut and debarked portions, infestation of stem-borer and leaf cutting weevil insect pests, and anthracnose disease on young leaves are the major pests which cause damage to the pruned trees.

As discussed earlier, to check microbial infection, paste of copper oxychloride must be applied on the cut surfaces immediately after pruning. Application of paste must be repeated wherever cut surfaces are seen exposed or the earlier layer of pasted is thinning. Biodynamic tree paste (BTP) developed at the Institute is an effective and economic alternative to copper oxychloride paste. BTP should be applied on the cut surfaces of pruned branches, injured and debarked portions, and sites of gum oozing and stem borer infestation. Application of BTP on entire trunk and branches of pruned tree is helpful in checking the incidences of stem borer insect pest and oozing of gum.

Infestation of stem-borer can be easily identified by wooden frass fallen on ground from the affected branches. Holes and oozing of gum in affected branches are the other indicators of its infestation. Larvae of the insect tunnel inside the trunk and destroy the conductive tissues. As a result, the branch and foliage start drying. Larvae can be pulled out of the hole by using thin wire or cycle spoke or they can be traced along the tunnel in the branch made by them from gum oozing spot. Larvae hidden inside the branch and trunk can be effectively controlled by placing cotton wick, soaked with Nuvan insecticide, inside the hole and sealing it with mud.

New shoots are affected with leaf cutting weevil insect pest and anthracnose disease. Leaf cutting weevil damages shoots by cutting the leaf across the lamina like scissors. It can be managed by two sprays of 0.2 % carbaryl (Sevin) insecticide (@ 3g per litre water) at an interval of 15 days. Brown spots on young leaves are the characteristic symptom of anthracnose disease. Copper oxychloride (3g per litre water) should be sprayed twice at an interval of 15 days for its management.

Preparation of Biodynamic Tree Paste (BTP):

Biodynamic Tree Paste (BTP) can be easily prepared in the orchard. Firstly, 40 kg each of cow dung, basalt powder or orchard soil and sand are thoroughly mixed like kneading of wheat flour. Later 250 g of cow horn manure (B.D. 500) and water is added to the above mixture and thus tree paste is ready for use.

xiii. Adventitious method of feeding

Old trees with weak growth can be invigorated by infusing the sap of younger seedlings into them. Several seedlings are grown close to the trunk of the tree. When they attain an age of two or three years, they are headed back to the height of 2′ to 3′ from the ground. The cut ends are shaped to a wedge form up to a length of about 2″ and are inserted into the bark of the tree. If needed both surfaces might be nailed and would be finally covered by grafting wax followed by firm tying with tape. If the tree is lacking in vigour due to unsatisfactory rootstock, the seedlings should be grafted into the scion not into the rootstock. In course of time, the seedlings get united to the tree and serve as its feeder. This could be practiced in mango, citrus, apple and in many other fruit plants. This method may also be followed when the collar region has been damaged.

xiv. Repairing of wounds

Any wound on the tree if allowed to remain exposed may attract the organisms of diseases from the surrounding atmosphere. So, they should be properly treated to encourage healing. If wound is small, simply painting with colour or any other disinfectant may suffice the purpose. In case of bigger wounds on the trunk, a special method of grafting called ‘bridge grafting’ is followed. Big hollows may be strengthened by scrapping off the inside diseased or rotten parts, smearing the exposed portion with coal tar and filling them with bricks and *kankar*. These are finally plastered with cement.

xv. Top working and frame working

It is a technique or method of rejuvenation wherein the objective is to upgrade seedling plantations of inferior cultivars with superior commercial cultivars or hybrids suitable for domestic or export market or the desired cultivar of the grower. The technique involves grafting with procured scions of desired cultivar on shoots emerged on pruned branches by adopting softwood grafting during monsoon season (Season of top working slightly varies from species to species as it also depends on availability of good shoot and scions). The scion shoots and the emerged shoots should be of same thickness.

Advantages

1. Increase the tree productivity /orchard productivity.
2. Conversion of old and senile orchards into productive orchards.
3. Conversion of seedling or inferior cultivar plantation /orchard into new orchard with desirable cultivar or cultivars through top working.
4. Possibility of grafting several cultivars on the same plant.
5. Increasing the fruit set of orchards by grafting few shoots with polliniser cultivars.
6. Additional income by selling the pruned wood during non-bearing season or period.

Disadvantages

1. Chances of death of plant if not done properly or on severe pruning.
2. Need good management in post-pruning period.
3. Loss of crop for 2-3 years.
4. Chances of pest and disease occurrence (stem borer, anthracnose etc.).
5. Needs skilled labour for thinning of shoots, removal of side shoots etc.
6. Top working technique can be successfully followed in crops like mango, sapota, aonla, cashew, guava, tamarind, jackfruit, etc.

Some practical hints for top working of these fruit plants are given below

Selection of plants to be rejuvenated

The plants, which are giving low yields of inferior quality even under proper management, should be selected. However, there is no limit to the age of the plants, provided they are healthy and vigorous.

Selection of mother plants for bud/scion wood

The plants of known history should be selected as mother plants. The plants giving outstanding yield of superior quality should be preferred. The mother plants should be free from major insect-pests and diseases. After the selection of mother plants, they should be maintained properly by following adequate management practices.

Methods for top working

Different methods can be employed while top working the fruit plants. However, the success is better with crown grafting, side veneer grafting and budding. These methods can be used singly or in combination.

Heading back of the plants to be rejuvenated

For top working, the plants to be rejuvenated are to be headed back. The height at which the plants to be headed back, depends on the age of the plant. In well protected orchard, where stray animals (cattle, goat and sheep etc.). are unable to cause any damage, the small plants in pre-bearing age may be cut at a height of nearly 4 to 5 feet from the ground level. In old and vigorous fruiting plants, it is desirable to cut the branches from the places where the plant is 6 to 9 inches in diameter. Thus, the height at which the branches should be cut would depend on the size of the plant. On the fruiting plants, it is very necessary to retain some lateral branches near to the ground to serve as nurse limbs and to provide shade to the trunk. These branches can be removed in the following years when enough crowns would develop by the new scion cultivar or cultivars used for top working.

Selection and preparation of scion/bud wood

For crown and side veneer grafting, shoots of 6 to 9 inches are preferred. They should be taken from the terminal mature shoots or current year's growth. If the length of the current year's growth is not sufficient, the previous year's growth may be used. However, for budding, only current year's growth should be utilized. Such shoots are normally available on young vigorous plants. The new growth can also be induced in old and devitalized plants by pruning one or more branches a few months before the actual requirement. Leaves from the selected shoots should be removed by secateurs about 7 to 10 days before the actual operation of budding/grafting, leaving the petioles attached to the shoots. During this period, the petioles that are left behind drop down, thus completely defoliated shoots are ready for use. Only those shoots are used for grafting in which the apical buds have swollen but have not actually started growing. If the apical buds are found in an active state of growth, the shoots must be discarded. Such shoots can, however, be used for taking buds for budding purpose.

Best time for budding and grafting

The crown grafting and side veneer grafting can be done during the rainy season, when there is plenty of humidity in the atmosphere and the weather is cloudy. Budding is successful when the maximum temperature varies between

90°F and 100°F. Thus, spring in plains and the summer months in the hills are best suited for budding. In the plains, the month of June has also been found to be suitable if there are good pre-monsoon showers.

Operations of budding and grafting

Crown grafting must be done immediately after the trunk or branch is cut. With the point of strong knife, two vertical parallel cuts at a distance of ½" to 1", starting from the cut surface downwards to a length of 3 to 4 inches, should be made so that bark is completely cut. By inserting the ebony end of a budding knife or some other pointed material between the bark and the wood, the bark should be separated from the wood and kept back in its position. Two slanting cuts are then made in the bud wood, one nearly 3 to 4 inches in length and the other nearly ½ to 1", to give it a wedge shape. It is important that the longer cut of 3 to 4 inches is given in one stroke with a sharp strong knife. The wedge is then inserted between the bark and the wood, the longer surface pointing inwards. Depending on the thickness of the trunk of branch, 2-3 or more grafts can be made in this way. These grafts should be tied firmly but not very tightly with a string used for stitching gunny bags. The cut surface should be covered with grafting wax (prepared by boiling together three parts resin, two parts bee wax and one-part coconut or linseed oil). Grafting, two parallel vertical cuts in the bark are joined by a third horizontal cut on the upper side. To facilitate insertion of ebony end of the knife or other pointed material for lifting the bark in thick branches and trunks, it is necessary to remove a triangular patch of the bark with the horizontal cut acting as a base. This triangular patch can be put back in its position after inserting the wedge-shaped portion of the bud wood between the bark and the wood. Budding is done on new shoots that grow from the pollarded branches but budding has to be done only on some selected ones, which should be properly spaced. Remaining shoots should be rubbed off as soon as they appear. Shield-shaped buds without wood are removed from the defoliated shoots and are inserted between the bark and the wood by making only on vertical cut on the bark of the shoot and loosening the bark with the ebony end of the budding knife. Budding tape prepared by passing strips of cloth through a hot mixture of the melted paraffin and resin (mixed in the preparation of 3:1) is used for tying the buds. To get best results, bud sticks should be used soon after they are removed from the mother plants. They can, however, be stored for 4 to 5 days in damp gunny pieces or in polythene film.

After care of the top worked tree

Crown and side grafts normally grow very vigorously. They should be protected from being blown down by storm by giving supports. The string used for tying should be cut and removed when it is found that it is girdling the graft union. During summer months, a shade must be provided on the southern-western side to protect the bark from scalding. Watch should be kept for the stem borer, presence of which can be detected by noticing saw dust like powder on the ground. Grubs should be traced by peeling off the bark, which is easily removed and killed. If holes are found in the wood, it means that the grubs have entered the wood. Diesel oil or petrol mixture can be used for destroying them. When thick branches are cut, it makes many years for the new bark to cover the cut surface. The central portion of the wood starts rotting, making trunks and branches hollow, and thus shortening the life of the plants. To prevent this, the cut surfaces should be covered during rainy season with a plaster, prepared by mixing 3 parts of sand and 1 part of the cement. Similarly, if the cut portions are plastered with Bordeaux mixture, it also helps in preventing the infection of the branches.

In frame working, only the smaller branches and shoots are replaced by scion of desired cultivar. The frame working is not successful in tropical and sub-tropical fruit plants, while temperate fruit plants are successfully frame worked by stub, awl and inverted 'L' method of grafting.

xvi. Wind breaks and fencing

Wind breaks are necessary for reducing the force and adverse effects of winds. The most effective are the double rows of tall trees, alternatively placed. Trees like sheesham, carambola, jamun, semal, paper mulberry and *Terminalis arjuna* can be effectively used for this purpose. Orchard area should be fenced with barbed wire along with suitable protective and economical hedge.

9.4 When Not to Rejuvenate

Before an attempt is made to rejuvenate the trees, the future income from the orchard should be taken into consideration. Sometimes it might be that the condition of the orchard is so poor that it is better to remove the old plantation and replace it by new one. It should be emphasized finally that trees during their first bearing years are not in need of re-invigoration even if they might be uneconomic. Satisfactory growth should be maintained by other cultural practices such as proper soil management, provision of adequate fertilizers and a water supply.

9.5 Rejuvenation of Old/Unproductive Mango Orchards

9.5.1 Glimpse of a Mango Orchard

In mango, 40-45 years old trees exhibit decline in fruit yield because of dense and overcrowded canopy. These trees do not get proper sunlight resulting in decreased production of shoots. New emerging shoots are weak and are unsuitable for flowering and fruiting. The population of insect-pests and the incidence of diseases increase in such orchards.

9.5.2 Rejuvenation Technique

The unproductive trees can be converted into productive ones by pruning them. Intermingled, diseased and dead branches are removed. Thereafter, undesirable branches of unproductive trees are marked. At the end of December, these marked branches are beheaded at 1.5 to 2.0 metres from distal end and the cut portions are pasted with copper oxychloride solution. During March-April, a number of new shoots emerged around the cut portions of the pruned branches. Only 8 to 10 healthy and outward growing shoots are retained at proper distance so that a good frame-work is developed in the following years. These rejuvenated trees are fertilized with 2.5 kg urea, 3.0 kg single super phosphate and 1.5 kg M.O.P./plant. The half dose of fertilizers is applied in the month of February and the other half at the end of June. The plants are irrigated at an interval of 15 days especially in the months of April, May and June for healthy growth of new shoots. In the first week of July, 150 kg of compost per tree is also applied. Unwanted emerging new shoots are regularly removed to maintain the tree canopy. It also helps in getting proper nourishment to retained shoots. After two years of pruning new shoots come into bearing and the yield of fruit increases gradually. Thus, old and unproductive trees are converted into productive ones.

9.5.3 Calendar of Activities for The First Year

December- January

- Marking of trees and their undesired branches for pruning.
- Pruning of marked branches in December.
- Pruning to be initiated from lower surface of the branches and later from upper surface to avoid cracking of branch and bark splitting.
- Application of copper oxychloride paste or biodynamic tree paste on the trunk, branches as well as cut surfaces to check microbial infection
- Ploughing and weeding in orchards in January.
- Preparation of basins and irrigation channels.

February- March

- Application of recommended full dose of single super phosphate (3.00 kg/tree) and muriate of potash 1.5 kg/tree and half dose of urea (1.25 kg/ tree) in basins in the end of February.

- Careful observation for infestation of stem borer insect pest in pruned trees. Upon identification of infestation, placing cotton wick soaked with nuvan in the holes followed by sealing with mud.
- Irrigation as per requirement during March.

April-May

- Irrigation as per requirement.
- Mulching in basins around trees.
- Hoeing and weeding in basins
- Care for new emerging shoots.
- Observation for incidence of stem-borer and its management.

June- July

- Thinning out undesired shoots while retaining about 8-10 healthy shoots with outwardly growth per pruned branch during June followed by spray of copper oxychloride
- Irrigation at an interval of 10-15 days.
- Application of remaining half does of urea i.e. 1.25 kg/tree during June.
- Application of FYM (120 kg/tree) in basins during July.
- Management of stem borer as described before.
- Spray of Copper oxychloride (3g/litre water) twice at an interval of 15 days if there is infestation of anthracnose on new leaves.
- It there is serious incidence of leaf cutting weevil, two sprays of 2% carbaryl (Sevin) @ 2 g/litre of water at an interval of 15 days may be done.

August- September

- Thinning out undesired shoots.
- Observation for incidence of stem-borer insect pest and anthracnose disease and their management.

October- November

- Cultural operations of ploughing, hoeing and weeding etc.
- Removal of dried and diseased twigs.
- Management of insect pests and diseases.
- Foliar spray of 2% urea during October for healthy vegetative growth.
- Marking of trees for pruning.

Proper nutrition, irrigation and plant protection must be ensured in subsequent years for successful rejuvenation.

9.6 Rejuvenation of Old/Unproductive Guava Orchards

9.6.1 Glimpse of a Guava Orchard

A procedure to rejuvenate and restore the production potential of old unproductive and wilt affected orchards has been developed, which employs pruning of branches at different periodicity and at different severities. Crowding and encroachment of guava trees with subsequent inefficient light utilization is an obvious problem with older orchards, if trees are not well managed. The internal bearing capacity of guava trees also decreases with time, due to overshadowing of internal bearing wood.

9.6.2 Rejuvenation Technique

The rejuvenation technology involves cutting of exhausted trees (showing marked decline in annual production) to the extent of 1.0 to 1.5 metre height above the ground level during May with the objective of facilitating new shoots. The newly emerging shoots are allowed to grow up to a length of about 40 to 50 cm which could be attained in 4-5 months of pruning.

These shoots are further pruned out to about 50% of its total length in October to facilitate emergence of multiple shoots below the pruning point. Profusely emerging shoots in the inner canopy are also pruned out to promote branching. The multiple shoots developed as a result of October pruning are capable of producing flower buds for the rainy season crop.

Those farmers keen to harvest the rainy season crop can allow the shoots to bear buds and fruits. However, as the winter crop has more marketing edge and value due to quality with the onset of rainy crop, shoot pruning (50%) is done again in May. This procedure of sequential and periodic pruning is continued every year for proper shaping of tree canopy and to ensure enhanced production of quality fruits during winter season.

References

Bose TK, Mitra SK and Sanyal D (2001). *Fruits: Tropical and Subtropical*, Naya Udyog Publishers, Calcutta.

Bose TK, Sanyal D and Sadhu MK (1998). *Propagation of Horticultural Crops*. Naya Prokash Publishers, Calcutta.

Brison FR and Adriance GW (2000). *Propagation of Horticultural Plants*. Biotech Books, New Delhi.

Chattopadhyay TK (1999). *A Textbook of Pomology*. Vol. I: *Fundamentals of Fruit Production*. Kalyani publishers, Ludhiana, India.

Sharma RR and Srivastava M (2004). *Plant Propagation and Nursery Management*. International Book Distributing Co. Lucknow.
Sharma RR (2002). *Propagation of Horticulture Crops (Principles and Practices)*. Kalyani Publishers, New Delhi.

e-resources

Bulletin on Rejuvenation of Old and Senile Mango Orchards, Central Institute of Subtropical Horticulture (ICAR), Rehmankhera, Lucknow, U.P.
e-Learning portal on Agricultural education, *https://ecourses.icar.gov.in.*
Rejuvenation of Uneconomic Orchards, Krishi Sandesh, *https://www.krishisandesh .com*

Outcomes Assessment

Part A

Answer the following questions true or false.

1. Crown and side grafts normally grow very vigorously. True/False
2. Old trees with weak growth cannot be invigorated by infusing the sap of younger seedlings. True/False
3. Biodynamic Tree Paste (BTP) can be easily prepared in the orchard. True/False
4. Drainage is not so important as water supply. True/False
5. Any good plants may be selected as mother plants. True/False

Part B

Answer the following questions.

1. What is rejuvenation?
2. The disbalanced root-shoot ratio can be corrected by judicious —.
3. Why do 40-45 years old mango trees exhibit decline in fruit yield?
4. Why are wind breaks necessary?
5. Crown grafting must be done immediately/later after the trunk or branch is cut.

Part C

Write a brief note on each of the following.

1. After-care of the top-worked tree.
2. Thinning of newly emerging shoots.
3. Overcrowding and unsystematic planting.
4. Rejuvenation of old/unproductive guava orchards.
5. Repairing of wounds in trees.

10

Integrated Nutrient Management

10.1 Introduction

- In case of integrated nutrient management, all types of nutrients are supplied in the form of organic or inorganic manures and fertilizers to maintain uniform soil fertility.
- Continuous use of fertilizers spoils the soil structure and soil becomes unproductive in due course.
- Fertilizers applied along with sufficient quantities of organic manures will improve the efficiency of the fertilizers. Green manuring should be done every year or once in two years during the rainy season.
- Suitable bio-fertilizers should also be applied from time to time.
- Plants need adequate supply of sixteen essential nutrients for their normal growth and production, out of which carbon (C), hydrogen (H), oxygen (O), nitrogen (N), phosphorus (P), potassium (K), calcium (Ca), magnesium (Mg) are needed in larger quantities and are referred as macro-nutrients, whereas, zinc (Zn), copper (Cu), iron (Fe), manganese (Mn), boron (B), molybdenum (Mo) and chlorine (Cl) are required in smaller quantities and thus called as micro-nutrients.
- An essential element is distinguished from a non-essential one on the basis of following criteria:
  i. The plant is unable to grow normally and complete its life cycle in the absence of the element.
  ii. The function of the element is specific and cannot be replaced by any other element.
  iii. The element plays a direct role in the metabolism of the plant.
- Carbon, hydrogen and oxygen are taken up by the plants from water and air; all other nutrients are drawn by plants from the soil.
- These elements are found in the soil in varying quantities depending upon the fertility of the soil. Fruit trees remove a large quantity of nutrient elements from the soil.

- Doses of nutrients vary from crop to crop. For example, nitrogen requirement per hectare basis varies from 50 kg for mango to 650 kg for grapes, whereas P_2O_5 requirement varies from 30 kg for mango to 434 kg for papaya.
- Potassium requirements of fruit crops are much higher than that of nitrogen or phosphorus. Potassium is very important for fruit quality.
- Even fertile soils have a limited ability to cope with the nutritional requirement of plants.
- For healthy growth and optimal yield, nutrients must be available to plants in correct quantity, proportion and in a usable form at right time. To fulfil these requirements chemical fertilizers or natural manures are needed.
- Fruit crops are mostly perennial and therefore, require long term nutrient management strategies.
- Single source of plant nutrients, *viz*., fertilizers, organic manures, FYM, bio-fertilizers, green manure is capable to satisfy the crop requirements.
- Application of chemical fertilizers though, makes the soil fertile and increases productivity of plants, but also has an adverse effect on soil and environment.
- Organic sources have enormous abilities to improve physical properties of soil, but have meagre nutrient contents to meet nutrient needs of fruit trees, when used alone.
- It is important that fertility and productivity of the soil be restored/ maintained on sustainable basis by combining organic and inorganic fertilizers.
- Use of organic manures, apart from improving physical and biological properties of soil, helps in improving the use efficiency of chemical fertilizers.

Therefore, there is a need to organize the supply of nutrients to crop through organic and renewable sources and strengthen the idea of 'Integrated Nutrient Management' (INM).

10.2 Definition

INM has been defined by different researchers as follows:

INM is defined as maintenance or adjustment of soil fertility and supply of plant nutrient to an optimum level for sustaining the desired crop productivity through optimization of benefit from all possible resources of plant nutrients in an integrated manner. INM is used to maintain or adjust soil fertility and plant nutrient supply to achieve a given level of crop production. This is done by optimizing the benefits from all possible sources of plant nutrients.

INM is the intelligent and combined use of inorganic, organic and biological resources so as to sustain optimum yields, improve or maintain the soil chemical and physical properties and provide crop nutrition packages which are technically sound, economically attractive, practically feasible and environmentally safe. The principal aim of the integrated approach is to utilize all the possible sources of plant nutrition in a judicious and efficient manner.

10.3 Concept of INM

The basic concept of INM is the promotion and maintenance of soil fertility for sustaining crop productivity through optimum use of all possible sources of nutrients like organic, inorganic and biological in an integrated manner appropriate to each farming situation. Improvement of soil fertility and productivity on sustainable basis through appropriate use of fertilizers and organic manures is the key principle and their scientific management for optimum growth and yield of crops in a specific agro-ecological condition is essential.

INM uses five major sub-concepts

1. Plant nutrients stored in the soil.
2. Plant nutrients, those present in the crop residues, organic manure and domestic wastes.
3. Plant nutrients purchased or obtained from outside the farm.
4. Plant nutrient losses, e.g., those removed from the field in crop harvest and lost from the soil through volatilization (ammonia and nitrogen oxide gases and leaching (nitrate, sulphate etc.)
5. Plant nutrient outputs, e.g., nutrient uptake by the crops at harvest time.

10.4 Needs for INM

- Unbalanced use of N: P: K have caused detrimental long-term effects on soil fertility.
- In areas subjected to intensive cultivation, application of mere chemicals is not sufficient for sustaining the yields, and it also leads to deficiency in the soil of secondary nutrients and micronutrients which limit crop productivity.
- Use of organic manure, crop residue and biodegradable rural and urban waste not only supplement the chemical fertilizers but also increase the

efficiency in nutrient supply, leading to improvement of physical and biological properties of the soil.

10.5 Importance of INM Systems

- The INM helps to restore and sustain soil fertility and crop productivity.
- It may also help to check the emerging deficiency of nutrients other than NPK.
- It brings economy and efficiency in fertilizer use and favourably affects the physical, chemical and biological environment of soil.
- It helps to produce fruits of high nutritional quality in sufficient quantity.
- Currently, decreasing soil fertility has also raised concerns about the sustainability of agricultural production.
- Future strategies for increasing agricultural productivity will have to focus on using available nutrient resources more efficiently, effectively than in the past.
- Integrated management of the nutrients is needed for proper plant growth, together with effective crop, water, soil, and land management.
- Restoration and sustenance of soil fertility and crop productivity.
- Economizing in fertilizer use and improvement in nutrient use efficiency.
- Favourable effect on the physical, chemical and biological health of soils.

10.6 Main Objectives of INM

- To reduce the dependence on chemical fertilizers.
- To maintain productivity on sustainable basis without affecting soil health.
- To conserve locally available resources and utilize them judiciously.
- To reduce the gap between nutrients used and nutrients harvested by the crop.
- To improve physical, chemical and biological properties of soil.
- To make soil healthy by providing balanced nutrients through different nutrient sources.
- To overcome or reduce the ill effects of continuous use of only inorganic chemical fertilizers.
- To improve economic status of farmers.
- To increase the fertilizer use efficiency (FUE).

- To ensure productive and sustainable agriculture.
- To reduce expenditure on costs of purchased inputs by using farm manure and crop residue, etc.
- To utilize the potential benefits of green manures, leguminous crops and bio-fertilizers.
- To prevent degradation of the environment.
- To meet the social and economic aspiration of the farmers without harming the natural resource base of agricultural production.

10.7 principles of INM

INM is based on three principles:

1. Assessment of basic soil fertility and climate.
2. Nature of the crop, not in isolation but as a part of the cropping system and yield targets.
3. At least 30% of the total nutrient levels of the NPK to be in organic form. This helps to estimate the fertilizer level, time of application to crops.

10.8 Components of INM

INM mainly emphasizes the integrated use of all the essential nutrients from different sources like chemical fertilizers, organic manures, green manures, bio-fertilizers, legume crops, locally available plant resources in a balanced proportion for sustainable soil health and productivity. Following are the different components of INM:

1. Use of inorganic fertilizers
2. Use of organic manures/ materials
3. Use of biological sources/bio-fertilizers
4. Maintaining the physical properties of soil
5. Management of problematic soils
6. Better/Judicious water management practices

10.8.1 Use of Inorganic Fertilizers

Fertilisers continue to be the most important ingredient of INM. The dependence on fertilisers has been increasing constantly because of the need to supply large amounts of nutrients in intensive cropping with high productivity.

The N: P_2O_5: K_2O use ratio is quite wide whereas application of K, S and micronutrients is often ignored. Domestic fertilizer production is inadequate to meet the requirements and the situation is not likely to improve in the near future. On the other hand, constraints like global price hike of fertilisers and raw materials would not permit fertiliser import in large quantities leading to a big gap between fertiliser supply and consumption. While organics and bio-fertilisers are expected to bridge a part of this gap, efficient use of fertilisers in narrowing the nutrient supply gap also needs greater emphasis.

Utilization of fertiliser nutrients by the crops vary from 30-50% in case of N, 15-20% in case of P and less than 5% in case of micronutrients. Thus, they are very important for sustaining and increasing food production. Different kinds of fertilizers are commercially available in the market for all the major and micronutrients. However, they are costly inputs and their excessive use may deteriorate the soil quality and food quality. Hence, there is a need to improve their use efficiency through efficient and balanced fertilizer management and essentially follow the four R's formula for judicious and effective nutrient/fertilizer management. They are:

- Right Type of fertilizers.
- Right Dose of fertilizers.
- Right Method of application.
- Right Time of application.

The following practices can be adopted for increasing the fertilizer use efficiency:

i. Incorporation of basal dose of fertilizers.
ii. The inorganic fertilizers should be applied at optimum soil moisture content.
iii. Split application of fertilizers at important phenological stages of crop growth.
iv. Use of coated and slow release fertilizers.
v. Foliar spray of micronutrients to overcome the nutrient deficiencies.
vi. Application of enriched FYM to avoid fixation of phosphorus.

Factors affecting the efficient use of fertilizers

1. Soil Factors

i. Soil texture
ii. Total nutrient content of soil
iii. Availability of nutrients
iv. Soil reaction

v. Soil erosion
vi. Drainage
vii. Soil management practices
viii. Presence of impermeable layer in the soil

2. Climatic and Environmental Factors

i. Temperature
ii. Relative humidity
iii. Rainfall
iv. Evaporation
v. Length of growing season/period
vi. Intensity of sunlight
vii. Duration of sunlight

3. Crop Factor

i. Cropping system
ii. Nutrient requirement of crops
iii. Crop adaptability to soil factors
iv. Crop production techniques
v. Crop residues

4. Systems of Farming

i. Wetland farming
ii. Garden land or irrigated upland farming
iii. Dryland or upland farming

5. Fertilizer Characters

i. Nature of fertilizers
ii. Solubility of fertilizers
iii. Dosage of fertilizers
iv. Method of application of fertilizers

10.8.2 Use of Organic Manures/Organic Materials

Manures are the organic materials derived from animal, human and plant residues which contain plant nutrients in complex organic forms. Manures with low nutrient content per unit quantity have longer residual effect besides improving soil physical properties compared to fertilizer with high nutrient content. Organic manures not only supply macro- and micro-nutrients, but also help improving the physical, chemical and biological properties of the soils. These manures, besides supplying nutrients to the first crop, also leave substantial residual effect on succeeding crops in the system.

Application of 30 to 50% of total nutrients in the form of organics would improve the physical properties of soil, provide congenial conditions for microbial activity, enhance the response of added nutrients, facilitate slow release of nutrients, reduce nutrient losses and provide a long-term build-up of soil fertility.

Thus, organic manures are of paramount importance not only in augmenting the crop production but also for making the agriculture more sustainable as an eco-friendly means of soil health management. Unlike inorganic fertilizers that supply only the major nutrients, organic manures serve as a store house of several plant nutrients and act as a good soil conditioner. Major sources of manures are:

1. Compost and FYM have traditionally been the important manures for maintaining soil fertility and ensuring yield stability. Cattle shed wastes—dung, urine and slurry from biogas plants.
2. Human habitation wastes—night soil, human urine, town refuse, sewage, sludge and sullage.
3. Poultry litter, droppings of sheep and goat.
4. Slaughter house wastes—bone meal, meat meal, blood meal, horn and hoof meal, fish wastes.
5. By-products of agro-industries—oil cakes, bagasse and press mud from sugar industry, fruit and vegetable processing wastes, spent-wash from distillery, molasses.
6. Crop wastes—sugarcane trash, stubbles and other related material.
7. Water hyacinth, weeds and tank silt.
8. Green manure crops and green leaf manuring material.
9. Sulphitation pressmud (SPM) has a great potential to supply nutrients in addition to favourable effects on soil properties. During the last three decades, SPM has assumed great importance as a nutrient supplement in sugarcane ratoon- wheat and other intensive cropping systems of the sugarcane growing areas.
10. Municipal solid wastes (MSW) and sewage-sludge are the other important nutrient sources available for integration with fertilizer inputs.

They are broadly classified as bulky organic manures, and concentrated organic manures.

10.8.2.1 Bulky Organic Manures

Bulky organic manures contain small percentage of nutrients and they are applied in large quantities. Farmyard manure (FYM), compost and green-manure are the most important and widely used bulky organic manures. Use of bulky organic manures has several advantages:

- They supply plant nutrients including micronutrients.
- They improve soil physical properties like structure, water holding capacity etc.
- They increase the availability of nutrients.
- Carbon dioxide released during decomposition acts as a CO_2 fertilizer.
- Plant parasitic nematodes and fungi are controlled to some extent by altering the balance of microorganisms in the soil.

Different bulky organic manures are discussed here:

i) Farm yard manure
ii) Sheep and Goat manure
iii) Poultry manure
iv) Compost
v) Night soil
vi) Sewage and sludge
vii) Crop residues
viii) Green manures
ix) Green Leaf manures
x) Legumes
xi) Sugarcane press mud
xii) Coir pith

i) Farm yard manure

Farmyard manure refers to the decomposed mixture of dung and urine of farm animals along with litter and left-over material from roughages or fodder fed to the cattle. On an average well decomposed farmyard manure contains 0.5 % N, 0.2 % P_2O_5 and 0.5 % K_2O.

All available litter and refuse are mixed with soil and spread in the shed so as to absorb urine. The next morning urine-soaked refuse along with dung is collected and placed in the trench. A section of the trench from one end should be taken up for filling with daily collection. When the section is filled up to a height of 45 cm to 60 cm above the ground level, the top of the heap is made into a dome and plastered with cow dung earth slurry. The process is continued and when the first trench is completely filled, second trench is prepared.

The manure becomes ready for use in about four to five months after plastering. Partially rotten farmyard manure has to be applied three to four weeks before sowing while well rotten manure can be applied immediately before sowing. Generally, 10 to 20 t/ha is applied, but more than 20 t/ha is applied to fodder grasses and vegetables. In such cases farmyard manure should be applied at least 15 days in advance to avoid immobilization of nitrogen. The existing practice of leaving manure in small heaps scattered in the field for a very long period leads to loss of nutrients. These losses can be reduced by spreading the manure and incorporating by ploughing immediately after application. Vegetable crops like potato, tomato, sweet-potato, carrot, radish, onion etc., respond well to the farmyard manure. The other responsive crops are sugarcane, rice, Napier grass and orchard crops.

ii) Sheep and goat manure

The droppings of sheep and goats contain higher nutrients than farmyard manure and compost. On an average, the manure contains 3% N, 1% P_2O_5 and 2% K_2O. It is applied to the field in two ways. The sweeping of sheep or goat sheds are placed in pits for decomposition and it is applied later to the field. The nutrients present in the urine are wasted in this method. The second method is sheep penning, wherein sheep and goats are kept overnight in the field and urine and faecal matter added to the soil is incorporated to a shallow depth by working blade harrow or cultivator.

iii)Poultry manure

The excreta of birds ferment very quickly. If left exposed, 50% of its nitrogen is lost within 30 days. Poultry manure contains higher nitrogen and phosphorus compared to other bulky organic manures. The average nutrient content is 3.03% N; 2.63% P_2O_5 and 1.4% K_2O.

iv) Compost

Compost is organic matter that has been decomposed in a process called composting. This process recycles various organic materials otherwise regarded as waste products and produces a soil conditioner. It is used in gardens, landscaping, horticulture, urban agriculture and organic farming. The compost itself is beneficial for the land in many ways, including as a soil conditioner, a fertilizer, addition of vital humus or humic acids, and as a natural pesticide for soil. In ecosystems, compost is useful for erosion control, land and stream reclamation, wetland construction, and as landfill cover.

At the simplest level, the process of composting requires making a heap of wet organic matter (also called green waste), such as leaves, grass, and food scraps, and waiting for the materials to break down into humus after a period of months. However, composting can also take place as a multi-step, closely monitored process with measured inputs of water, air, and carbon- and nitrogen-rich materials. The decomposition process is aided by shredding the plant matter, adding water and ensuring proper aeration by regularly turning the mixture when open piles or "windrows" are used. Fungi, earthworms and other detritivores further break up the material. Bacteria requiring oxygen to function (aerobic bacteria) and fungi manage the chemical process by converting the inputs into heat, carbon dioxide, and ammonium.

v) Night soil

Night soil is a historically used euphemism for human excreta collected from cesspools, privies, pail closets, pit latrines, privy middens, septic tanks, etc. This material was removed from the immediate area, usually at night, by workers employed in this trade. Sometimes it could be transported out of towns and sold on as a fertilizer. Human excreta may be attractive as fertilizer because of the high demand for fertilizer and the relative availability of the material to create night soil. In areas where native soil is of poor quality, the local population may weigh the risk of using night soil. The use of unprocessed human faeces as fertilizer is a risky practice as it may contain disease-causing pathogens. Nevertheless, in some developing nations it is still widespread. These risks are reduced by proper faecal sludge management, e.g., via composting. The safe reduction of human excreta into compost is possible. Some municipalities create compost from the sewage sludge, but then recommend that it only be used on flower beds, not vegetable gardens. Some claims have been made that this is dangerous or inappropriate without the expensive removal of heavy metals.

vi) Sewage sludge or Biosolids

The term "biosolids" recently has been introduced by the waste water treatment industry. The industry defines biosolids as sewage sludge that has undergone sufficient treatment for stabilization and pathogen reduction, and that is of sufficiently high quality to be used specifically for land application. The term "biosolids" also helps to distinguish sewage sludge from industrial sludge by emphasizing that the former is produced by a biological process.

vii) Crop residues

Crop residues have several competitive uses and may not be always available as an ingredient of INM. Yet in the regions like North-West India where mechanical harvesting is practiced, a sizeable quantity of residues is left in the field, which can form a part of nutrient supply. There are large amounts of residues of other crops like potato, sugarcane, vegetables, etc., which are practically wasted in most cases. Although cereal crop residues are valuable cattle-feed, they can be used to supplement the fertilisers wherever available in excess of the local needs.

Disposal of rice straw in Trans- and Upper Gangetic Plains has emerged as a great problem. In these combine-harvested areas farmers opt to burn the residues *in-situ*, losing precious nutrients on one hand and polluting environment on the other. Recycling of these residues back to fields helps to build stable organic matter in the soil, as also to sustain crop yield levels.

viii) Green manures

Green manuring is the practice of growing lush plants and ploughing *in-situ* when succulent to improve the soil properties. Green manures when incorporated into the soil, not only supply the nutrients, but also improve the crop yield by enhancing soil fertility, improving soil structure, preventing soil erosion, and also improves the microbial activity.

Green manure crops suitable for different ecosystems are given below:

a. Daincha
b. Manila agathi
c. Sithagathi
d. Sunhemp
e. Kozhingi
f. Pillipesara
g. Kalopagonium.

ix) Green leaf manuring

Green leaf manuring is the practice of collecting green leaves and tender green twigs from shrubs and trees grown on bunds and waste lands and applying to the main fields for improving the fertility status of soil. Some examples are:

a. Neem
b. Pungam

c. Glyricidia
d. Erukku, and
e. Subabul.

x) Legumes

Legumes have a long-standing history of being soil fertility restorers due to their ability to obtain N from the atmosphere in symbiosis with Rhizobia. Legumes could prove an important ingredient of INM when grown for grain or fodder in a cropping system, or when introduced for green manuring. Legumes grown as green manure, forage or grain crops improved the productivity of rice-wheat cropping system (RWCS) and rejuvenated soil fertility.

xi) Sugarcane press mud

Sugarcane is used in the preparation of jaggery. The spent material after processing is known as press mud. It contains high amounts of cellulose and hemicelluloses which needs to be broken down into simpler constituents. Hence, the raw material needs to be properly digested and composted before it is applied to the soil.

xii) Coir pith

This is the raw material obtained from coconut. It is available in plenty and also the cost is reasonable. It acts as excellent mulch in young plantations and conserves moisture during the drought months. It has the capacity to absorb moisture many times over its weight and is used in composting and nursery preparation. It has high amounts of silica and hence should be sparingly applied.

10.8.2.2 Concentrated Organic Manures

Concentrated organic manures have higher nutrient content than bulky organic manure. The important concentrated organic manures are oilcakes, blood meal, fish manure etc. Animal by-products are permitted in certified organic production as soil amendments. They are also known as organic nitrogen fertilizer. Before their organic nitrogen is used by the crops, it is converted through bacterial action into readily usable ammoniacal nitrogen and nitrate nitrogen. These organic fertilizers are, therefore, relatively slow acting, but they supply available nitrogen for a longer period. Examples are:

i) Oil cakes
ii) Blood meal
iii) Meat meal

iv) Fish meal

v) Bone meal

i) Oil cakes

After oil is extracted from oilseeds, the remaining solid portion is dried as cake which can be used as manure. The oil cakes are of two types: edible oil cakes and non-edible oil cakes. Edible oil cakes can be safely fed to livestock, e.g., groundnut cake, coconut cake etc., and non-edible oil cakes are not fit for feeding livestock, e.g., castor cake, neem cake, mahua cake etc., Both edible and non-edible oil cakes can be used as manures. However, edible oil cakes are fed to cattle and non-edible oil cakes are used as manures especially for horticultural crops. Nutrients present in oil cakes, after mineralization, are made available to crops 7 to 10 days after application. Oilcakes need to be well powdered before application for even distribution and quicker decomposition.

The average nutrient content of different oil-cakes is presented in the following table:

Oilcake	Nutrient content (%)		
Non-edible Oilcakes	N	P	K
Castor cake	4.3	1.8	1.3
Cotton seed cake (undecorticated)	3.9	1.8	1.6
Karanj cake	3.9	0.9	1.2
Mahua cake	2.5	0.8	1.2
Safflower cake (undecorticated)	4.9	1.4	1.2
Edible oil-cakes			
Coconut cake	3.0	1.9	1.8
Cotton seed cake (decorticated)	6.4	2.9	2.2
Groundnut cake	7.3	1.5	1.3
Linseed cake	4.9	1.4	1.3
Niger cake	4.7	1.8	1.3
Rape seed cake	5.2	1.8	1.2
Safflower cake (decorticated)	7.9	2.2	1.9
Sesamum cake	6.2	2.0	1.2

ii) Blood meal

Blood meal is a dry, inert powder made from blood, used as a high-nitrogen organic fertilizer and a high protein animal feed. It contains N = 13.25%, P = 1.0%, K = 0.6%. It is one of the highest non-synthetic sources of nitrogen. It usually comes from cattle or hogs as a slaughterhouse by-product. Blood meal is sometimes used as a composting activator.

iii) Meat meal

The meat of dead animals is dried and converted into meat meal which is a good source of nitrogen.

iv) Fish meal

Fishmeal is generally used to feed farm animals in agricultural setting. Because it is calorically dense and cheap to produce, fishmeal has played a critical role in the growth of factory farms and it is possible to breed and feed the number of farm animals. Fish meal contains N = 4 - 10 %, P = 3 - 9% and K= 0.3 - 1.5 % as organic manure.

v) Bone meal

The bone of dead animals is dried and converted into bone meal which is a good source of phosphorus.

Average nutrient content of animal based concentrated organic manures is given as follows:

Organic manures	Nutrient content (%)		
	N	P_2O_5	K_2O
Blood meal	10 - 12	1 - 2	1.0
Meat meal	10.5	2.5	0.5
Fish meal	4 - 10	3 - 9	0.3 - 1.5
Horn and Hoof meal	13	-	-
Raw bone meal	3 - 4	20 - 25	-
Steamed bone meal	1 - 2	25 - 30	-

10.8.3 Use of Biological Sources / Biofertilizers

Bio-sources are the specific microorganisms derived either from the nodules of plant roots or from the soil of root zone (rhizosphere). These microorganisms have the ability to fix atmospheric nitrogen either by living symbiotically with the roots of leguminous plants or non-symbiotically (free living). They transform atmospheric nutrients from non-usable to usable form through biological processes. There is a lot of potential of this source to supplement inorganic fertilizers for productive agriculture. Suitable biological sources for different ecosystem are given below:

Biofertilizers

Biofertilizers are the materials containing living or latent cells of agriculturally beneficial microorganisms that play an important role in improving soil fertility and crop productivity due to their capability to fix atmospheric N, solubilize/mobilize P and decompose farm waste resulting in the release of plant nutrients. The extent of benefit from these microorganisms depends on their number and efficiency which, however, is governed by a large number of soil and environmental factors. Bacterial cultures like *Rhizobium*, *Azospirillum* and *Azotobacter* have the ability to fix atmospheric N which in turn increase N-supply to the crops. Bacterial cultures of *Pseudomonas* and *Bacillus* species and fungal culture of *Aspergillus* species help to convert insoluble P into plant usable forms and thus improve phosphate availability to the crops. Similarly, fungi like Vesicular Arbuscular Mycorrhizae (VAM) increase nutrient uptake particularly that of P due to increased contact of roots with larger soil volume. *Rhizobium* is the most well-known bacterial species that acts as the primary symbiotic fixer of N. These bacteria can infect the roots of leguminous plants, leading to the formation of lumps or nodules where the N fixation takes place. The bacterium's enzyme system supplies a constant source of reduced N to the host plant and the plant furnishes nutrients and energy for the activities of the bacterium.

There are broadly seven types of bio-fertilizers:

i) Blue-green algae

- Blue-green algae or cyanobacteria are free-living nitrogen-fixing photosynthetic algae that are found in wet and marshy conditions.
- Blue-green algae are so named for their colour but they may also be purple, brown or red.
- They are easily prepared on the farm but can be used only for rice cultivation when the field is flooded and do not survive in acidic soils.

ii) Azolla

- Azolla is a free-floating water fern that fixes nitrogen in association with a specific species of cyanobacteria.
- Azolla is a renewable bio-fertilizer and can be mass-produced on the farm like blue-green algae. It is a good source of nitrogen and on decomposition, a source of various micronutrients as well.
- Its ability to multiply fast means it can stifle and control weeds in (flooded) rice fields.
- Azolla is also used as a green manure and a high-quality feed for cattle and poultry.

iii) Phosphate-breaking micro-organisms

- These are a group of bacteria and fungi, e.g., VAM capable of breaking down insoluble phosphates to make them available to crops.
- Their importance lies in the fact that barely one third of phosphorous in the soil is actually available to the crop as the rest is insoluble.
- They require sufficient organic matter in the soil to be of any great benefit.

iv) Rhizobium

- Suitable rhizobial strain is treated with pulses as seed inoculation helps to increase the grain yield by fixing nitrogen from the atmosphere.

v) Azospirillum

- It is treated with sorghum, cumbu, ragi, maize and cotton as seed inoculation which fixes appreciable quantities of atmospheric nitrogen through associative symbiosis.

vi) Azotobacter

- It is capable of fixing atmospheric nitrogen as free-living microorganisms.
- This should be treated with crops like rice, sorghum, ragi and maize thus saving 25 % of nitrogenous fertilizer.

vii) Mycorrhizae

- It is a symbiotic association of fungi with the roots of seed plants.
- The mobility of nutrients mainly phosphorus and also zinc and sulphur from the soil to the plants is mediated by this association.
- This association also produces growth promoting substances, which stimulate plant growth thus increasing the yield.

(a) Ectotrophic Mycorrhizae

The hyphae of fungi occur on the root surface and penetrate only the intercellular spaces.

(b) Endotrophic Mycorrhizae

The hyphae of fungi occur mainly within cells of the host plants.

10.8.4 Maintaining the Physical Properties of Soil

Physical properties such as soil aggregation, soil texture, structure, aeration, water holding capacity (WHC), infiltration rate, etc., should be maintained regularly through better cultivation practices and organic manure applications to maintain soil fertility and nutrient availability.

10.8.5 Management of Problematic Soils

Problematic soils such as acid soils, saline and alkaline soils, water logged soils are known to decrease the productivity of the soil. Acid soil is having the problems like toxicities of Fe, Mn, and Al, and also deficiency of P and Mo. Similarly, saline and alkali soils are showing the deficiency of Fe, Mn, Zn and Cu and also toxicities of Mo. These soils should be regularly managed and reclaimed through the application of soil amendments such as lime for acid soil, gypsum for alkali soils and other organic and inorganic materials based on soil test results. It helps to improve soil fertility and productivity and sustain the yield.

10.8.6 Better/Judicious Water Management Practices

Plants absorb the nutrients from the soil only in a dissolved state and sufficient moisture is therefore required for utilizing the nutrients of the soil. Management of moisture in the soil by improved and modern irrigation techniques like drip or sprinkler or basin where the rainfall is low and draining the soil where it is subjected to stagnation of water helps to increase water and nutrient availability to the crops.

10.9 Implementation of INM Activities

Different stages of implementation of INM are as follows:

1. Diagnosis phase: collection of background information.
2. Analysis of constraints.
3. Preparing potentiality and feasibility summary
4. On-farm demonstrations.
5. Evaluation of INM activities.

Diagnostic Phase of INM

In the first stage of diagnostic phase information with regard to the following is collected and analysed:

i. Farming/cropping systems.
ii. Crop varieties grown.

iii. Awareness about soil fertility problems.
iv. Use of chemical fertilizers, lime/dolomite and other agro-chemicals.
v. Use of organic manure.
vi. Availability of fertilizer and other inputs.
vii. Irrigation sources and practices.
viii. Soil testing service facility.
ix. Constraints in the adoption of INM technologies.
x. Consideration of market opportunities.

The INM technologies must be compatible with the local farming system if they are to find acceptance and adoption. Therefore, attention must be paid to examine the interaction among different components of INM and the management of crops and animals that form the farming system.

Some of the agronomic parameters which need attention are cropping pattern, intercropping practices, biological condition of the field (weeds, diseases, and insects), soil conditions, irrigation facilities and climatic conditions.

Common Constraints of INM

Common constraints encountered by the farmers in adoption of INM technology are as follows:

1. Non-availability of FYM.
2. Difficulties in growing green manure crops.
3. Non-availability of bio-fertilizers.
4. Non-availability of soil testing facilities.
5. High cost of chemical fertilizers.
6. Non-availability of water.
7. Lack of knowledge and poor advisory services.
8. Non-availability of improved seeds.

References

Aulakh MS and Grant CA (2008). *Integrated Nutrient Management for Sustainable Crop Production*. The Haworth Press, Traylor & Francis Group, New York.

Chand S (2019). *Integrated Nutrient Management for Sustainable Agriculture*. Indian Books and Periodicals, New Delhi.

Dwivedi BS, Singh VK and Dwivedi V (2004). Application of phosphate rock, with or without *Aspergillus awamori* inoculation to meet P demands of rice-wheat systems in the Indo-Gangetic plains of India. *Aus. J. Exp. Ag.*, **44**: 1041-1050.

Roy RN and Ange AL (1991). *Integrated Plant Nutrition System (IPNS) and Sustainable Agriculture*, pp. SV/1-1-SV/1-12, FAI, New Delhi.

Singh M, Dwivedi BS and Datta SP (2012). Integrated nutrient management for enhancing productivity, nutrient use efficiency and environmental quality. In: *Soil Science in the Service of Nation*, pp. 55-67, ISSS, New Delhi.

Yadav RL, Singh SR, Prasad K, Dwivedi BS, Batta RK, Singh AK, Patil NG and Chaudhary SK (2000). Management of irrigated agroecosystem. In: Yadav JSP and Singh GB (eds), Natural *Resource Management for Agricultural Production in India.* India Soc. Soil. Sci. Publ. pp. 775-870, New Delhi.

e-resources:

e-Learning portal on Agricultural education, *https://ecourses.icar.gov.in*

https://www.biotecharticle.com

https://megagriculture.gov.in

www.soilmanagementindia.com

Outcomes Assessment

Part A

Answer the following questions true or false.

1. Blue-green algae are free-living nitrogen-fixing photosynthetic algae. True/False
2. Plants absorb the nutrients from the soil only in a dissolved state. True/False
3. Concentrated organic manures have lower nutrient content than bulky organic manures ones. True/False
4. Azotobacter is capable of fixing atmospheric nitrogen as symbiosis. True/False
5. Bone meal is a good source of potassium. True/False

Part B

Answer the following questions.

1. Define biofertilizers.
2. Saline and alkali soils show the deficiency of Fe, —, — and Cu.
3. There are broadly seven/ten types of biofertilizers.
4. Castor cake is edible/non-edible.
5. Cite two common constraints of INM.

Part C

Write a brief note on each of the following.

1. Farm yard manure.
2. Biofertilizers.
3. Compost.
4. Main objectives of INM.
5. Concept of INM.

11

Integrated Pest Management

11.1 Introduction

- Pest problems in fruit production has increased the cost of tree/plant protection and squeezed the profit of the fruit growers.
- In India, there are several existing or potential key pests for each fruit trees whose population densities must be kept at sub-economic levels to sustain higher per unit production at the farmers' fields.
- During the second part of 20th century, the major emphasis was on pest control, which involved a set of actions to avoid, alternate or delay impact of pests on crops, which mainly involved pesticide application with the sole aim of eradication.
- Since horticultural crops like fruits are intensively cultivated that allow more inputs cost, the use of insecticides for the control of specific pests of a particular fruit crop has increased manifold during the last few decades. However, the toxic materials generated from chemical use in crop production has polluted the environment, because majority of them are not biologically degradable. They have harmed consumers' and farmers' health.
- Uses of broad-spectrum chemicals for pest control have a very negative impact on our efforts to conserve biodiversity. Because of their broad-spectrum nature of effect, it not only kills enemy insects but also useful insects like honey bees.
- The pest control option based on pesticide applications virtually eliminates the bee keeping activity.
- Now, a more environment friendly and economical alternative for fruit cultivation has been adopted, namely **'Integrated Pest Management (IPM)'**.
- Due to considerable awareness about the harmful effects of pesticides on human health and the environment, recently there has been a renewed interest in IPM in almost all countries.

11.2 Definitions

- Pest is any living organism, which competes with human beings for resources of its interest. A phytophagous insect assumes pest status when its population is alarming to inflict significant damage in yield or quality of any economic crop such as fruits.
- Integrated pest management can be defined as "a sustainable approach of management of pests by the combination of biological, cultural, mechanical and chemical tools in a way that minimize economic, health and environmental risks".
- In other words, IPM is economically justified and sustainable system for the protection of crops that leads to maximum fruit productivity without having any adverse effect on environment.

11.3 Objectives of IPM

- To eliminate or mitigate economic and health damage caused by pests
- To minimize the use of pesticides and the risk to human health and the environment associated with pesticide applications
- To use integrated methods, site or pest inspections, pest population monitoring, an evaluation of the need for pest control, and one or more pest control methods, including sanitation, structural repairs, mechanical and living biological controls, other non-chemical methods, and, if nontoxic options are unreasonable and have been exhausted, least toxic pesticides.
- To reduce the use of synthetic organic pesticides
- To adopt environmentally sound tactics for pest control
- Pest minimal risk of human health
- Re-usable return on investment
- To provide consumable safe food

11.4 The Six IPM Programme Essentials

1. **Monitoring:** This includes regular site inspections and trapping to determine the types and infestation levels of pests at each site. Regular observation is critically important. Visual inspection, insect and spore traps, and other methods are used to monitor pest levels.
2. **Record-Keeping:** A record-keeping system is essential to establish trends and patterns in pest outbreaks. Information recorded at every inspection or treatment should include pest identification, population size, distribution, recommendations for future prevention, and complete

information on the treatment action. Record-keeping is essential, as it is a thorough knowledge of target pest behaviour and reproductive cycles.

3. **Action Levels:** Pests are virtually never eradicated. An action level is the population size which requires remedial action for human health, economic, or aesthetic reasons.
4. **Prevention:** Preventive measures must be incorporated into the existing structures and designs for new structures. Prevention is and should be the primary means of pest control in an IPM programme.
5. **Tactics Criteria:** Under IPM, chemicals should be used only as a last resort only, but when used, the least-toxic materials should be chosen, and applied to minimize exposure to humans and all non-target organisms.
6. **Evaluation:** A regular evaluation programme is essential to determine the success of the pest management strategy

11.5 Stages in Crop Protection Leading to IPM

1	Subsistence phase	Only natural control, no insecticide use
2	Exploitation phase	Applying more pesticides, growing HY varieties and get more yield and returns
3	Crisis phase	Due over use pesticides, problem of resurgence, resistance, secondary pest outbreak, increase in production cost
4	Disaster phase	Due to increased pesticide use - No profit, high residue in soil - Collapse of control system
5	Integrated Management Phase	IPM integrates eco-friendly methods to optimize control rather than maximise it.

11.6 Tools or Components of Integrated Pest Management (arranged in increasing order of complexity)

1. Cultural practices
2. Mechanical practices
   2.1 Mechanical destruction
   2.2 Mechanical exclusions
   2.3 Appliances based on mechanical control methods
3. Biological practices
4. Use of Bio-pesticides
5. Use of Botanicals in pest control
6. Pheromones in pest management
7. Biotechnological approaches
8. Regulatory practices

9. Pest surveillance and monitoring
10. Chemical practices

11.6.1 Cultural Practices

Cultural methods of pest control consist of regular farm operations in such a way which either destroy the pests or prevent them from causing economic loss. The various cultural practices are as under:

a. Preparation of nurseries or main fields free from pest infestation by removing plant debris, trimming of bunds, treating of soil and deep summer ploughing which kills various stages of pests.
b. Testing of soil for nutrients deficiencies on the basis of which fertilizers should be applied.
c. Selection of clean and certified seeds and treating seeds with fungicide or bio-pesticides before sowing for seed borne disease control.
d. Selection of seeds of relatively pest resistant/tolerant cultivars which play a significant role in pest suppression.
e. Adjustment of time of sowing and harvesting to escape peak season of pest attack.
f. Rotation of crops with non-host crops. It helps in reduction of incidence of soil borne diseases.
g. Proper plant spacing which makes plants more healthy and less susceptible to pests.
h. Optimum use of fertilizer. Use of FYM and bio-fertilizers should be encouraged.
i. Proper water management. As the high moisture in soil for prolonged period is conducive for development of pests especially soil borne diseases.
j. Proper weed management. It is well-known fact that most of the weeds besides competing with crop for micro-nutrients also harbour many pests.
k. Setting up yellow pan sticky traps for white flies and aphids at far above canopy height.
l. Synchronized sowing. Here community approach is required to sow the crops simultaneously in vast area so that pest may not get different staged crops suitable for its population build up and if pest appears in damaging proportion, control operation could be applied effectively in whole area.

m. Growing trap crops on the borders or peripheries of fields. There are certain crops which are preferred more by a pest species are known as trap crops for that pest. By growing such crops on the border of the fields, pest population develop there which can be either killed by using pesticides or its natural enemies are allowed to develop there for natural control.

n. Inter-cropping or multiple cropping wherever possible. All the crops are not preferred by each pest species and certain crops act as repellents, thus keeping the pest species away from preferred crops resulting in reduction of pest incidence.

o. Harvesting as close as to ground level. This is because certain developmental stages of insect pests/diseases remain on the plant parts which act as primary inoculum for the next crop season. Hence, harvesting crops at ground level will lessen the incidence of pests in next season.

p. Before planting, nursery plants be sprayed/dipped in copper fungicide/ bio pesticide solutions to protect the plants from soil borne diseases.

q. While pruning fruit trees, remove crowded/dead/broken/diseased branches and destroy them. Do not pile them in the orchards which may act as source of pest infestation.

r. Large pruning wounds should be covered with Bordeaux paste/paint to protect the plants from pest/disease attack.

s. Keeping bee hives or placing flower bouquets of pollinizer cultivars facilitate better pollination and subsequent fruit set.

t. Selection of high yielding cultivars for different crops.

u. Selection of comparatively pest resistant/tolerant cultivars.

11.6.2 Mechanical Practices

a. Removal and destruction of egg masses, larvae, pupae and adults of insect pests and diseases parts of plants wherever possible.

b. Installation of bamboo cage cum bird perches in the field and placing parasitized egg masses inside them for conservation of natural enemies and withholding of pest species wherever possible.

c. Use of light traps and destruction of trapped insects.

d. Use of rope for dislodging leaf feeding larvae, e.g., caseworm and leaf folders.

e. Installation of bird scarers in the field where required.

f. Installation of bird perches in the field for allowing birds to sit and feed on insects and their immature stages, viz., eggs, larvae and pupae.
g. Use of pheromones for mating disruption and kill zone creation.
h. Use of pheromone traps for monitoring and suppression of pest population.
i. Use of pheromone traps for mass trapping.

The mechanical practices can be categorised as follows:

11.6.2 .1 Mechanical Destruction

a. Hand picking of caterpillars
b. Hooking of rhinoceros beetle adult with iron hook
c. Sieving and winnowing for stored product insect control
d. Shaking plants to dislodge caseworm in rice-to dislodge June beetles from neem trees

11.6.2.2 Mechanical Exclusion

a. Wrapping of fruits against pomegranate fruit borer
b. Banding with grease against mango mealy bug
c. Trenching for larvae of red hairy caterpillar
d. Tin barrier around coconut tree trunk to prevent rat damage
e. Rat proof structure in storage godowns

11.6.2.3 Appliances Based on Mechanical Control Method

a. Light trap
b. Yellow sticky traps - for attracting aphids and jassids
c. Bait trap - fish meal trap for sorghum shoot flies
d. Methyl eugenol trap - for fruit flies
e. Probe trap - for stored product insects
f. Pheromone trap - for various adult insects
g. TNAU automatic insect removal bin - for stored product insects

11.6.3 Biological Practices

a. Biological control of insect pests and diseases through biological means is the most important component of IPM.
b. In broader sense, biocontrol is the use of living organisms to control unwanted living organisms (pests).

c. In other words, deliberate use of parasitoids, predators and pathogens to maintain pest population at level below those causing economic loss either by introducing a new bioagent into the environment of pest or by increasing effectiveness of those already present in the field.

Parasitoids

These are the organisms which lay eggs in or on the bodies of their hosts and complete their life cycles on host bodies as a result of which hosts die. A parasitoid may be of different types depending on the host developmental stage in or on which it completes its life cycle. For example, egg, larval, pupal, adult, egg-larval and larval-pupal parasitoids. Examples are different species of *Trichogramma, Apanteles, Bracon, Chelonus, Brachemeria, Pseudogonotopus* etc.

Predators

These are free living organisms which prey upon other organisms for their food. Examples are different species of spiders, dragon flies, damsel flies, lady bird beetles, *Chrysopa* species, birds etc.

11.6.4 Bio-Pesticides

a. These are micro-organisms which infest and cause diseases in their hosts as a result of which hosts are killed.
b. Major groups of pathogens are fungi, viruses and bacteria. Some nematodes also cause diseases in some insect pests.
c. Important examples of fungi are different species of *Hirsutella, Beauveria, Nomurae* and *Metarhizium* which have been reported to infest and kill large number of insects (up to 90%) in the fields.
d. Among viruses, most important examples are of nuclear polyhedrosis virus (NPV) and granulosis viruses.
e. Outbreak of viruses in armyworms, cut worms, leaf folders, hairy caterpillars and plant hoppers have been reported many times.
f. Among bacteria, *Bacillus thuringiensis* (B.t.) and *B. popilliae* are very common examples.
g. Diseases of pests can be mass multiplied in the laboratory at a low cost in liquid or powdered formulations that can be sprayed like ordinary chemical pesticides. These formulations are known as bio-pesticides.

The different types of biocontrol practices are grouped as under

Introduction

In this process, a new species of bio-agent is introduced into a locality for its establishment against its host. This is done only after thorough laboratory examination and field trials for its efficacy.

Augmentation

In this process, the population of natural enemies already present in the area is increased by releasing either laboratory reared or field collected bio-agents of same species in such number as would require to suppress the pest population in that area.

Conservation

This is most important component of biological control and plays a major role in pest suppression. In this process, natural enemies present in the nature are protected from being killed.

The different practices required to protect the natural enemies are as below

- Collection of parasitised egg masses and placing them in bamboo cage-cum-bird perches for allowing emergence of parasitoids and withholding of pest larvae.
- Use of pheromone traps for mass trapping.
- Educating farmers through field days, radios and TV to differentiate pests and defenders and sparing the defenders during field sprays.
- Chemical spray should be adopted as last resort and that too after observing pest defender ratio.
- Use of broad-spectrum pesticides should be avoided.
- Only selective and relatively environmentally friendly (REF) pesticides should be used where necessary.
- As far as possible strip or spot application of pesticides be carried out.
- Adjustment of time of sowing and harvesting to avoid the peak season of pest attack.
- Growing trap crop on the borders of main fields before the actual sowing of crop to trap pest and develop natural enemies.
- Root dip/seedling treatment for gall midge prone area.
- Recommended dose and concentration of pesticides should be used.

11.6.5 Use of Botanicals in Pest Management

a. Neem oil at 2% and neem seed kernel extract (NSKE) at 5% with liquid soap 0.05% have been proven affective against major pests of rice, sucking pests of cotton and vegetable.

b. Neem cake applied at 250 kg/ha at last ploughing before sowing has been found effective against cotton stem weevil and soil insects of many other crops.

c. Neem seeds contain more than 100 compounds among which *azadirachtin* has been found to be biologically most active.

d. The biological effects of neem products are insect growth regulation, feeding deterrent and oviposition deterrent effect.

e. Commercial Neem formulations are available in market which contain varying levels of *azadirachtin* (from 0.03% to a maximum of 5%).

f. In India more than 50 firms are manufacturing neem formulations which are available in different brand names.

A few examples are given below:

Sl.No.	Brand name	Azadirachtin content
1.	Nimbecidine	0.03%
2.	Neem guard	0.03%
3.	Bioneem	0.03%
4.	Jaineem	0.03%
5.	Neem gold	0.15%
6.	Fortune-aza	0.15%
7.	Econeem	0.3%
8.	Achook	0.5%
9.	Neem azal TS	1.0%
10.	Neem azal F	5.0%

11.6.6 Pheromones in Pest Management

a. Pheromones are chemical substances released by insects which attract other individuals of the same species.

b. Sex pheromones have been used in pest management in the following ways

   a. Monitoring

   b. Mating disruption

   c. Mass trapping

c. These methods can be successfully included in organic method of pest management.

Sex pheromones of the following insects are commercially available in the market:

Commercially available sex pheromones for insects

Sl. No.	Common Name	Scientific name
1.	American bollworm	*Helicoverpa armigera*
2.	Pink bollworm	*Pectinophora gossypiella*
3.	Spotted bollworm	*Earias vitella*
4.	Spiny bollworm	*Earias insulana*
5.	Tobacco cutworm	*Spodoptera litura*
6.	Early shoot borer of sugarcane	*Chilo infuscatellus*
7.	Yellow stem borer of rice	*Scirpophaga incertulas*
8.	Diamond back moth	*Plutella xylostella*
9.	Mango fruit fly	*Bactrocera dorsalis*
10.	Melon fruit fly	*Bactrocera cucurbitae*

d. Aggregation pheromones of red palm weevil and *Rhinoceros* beetle of coconut are also available in market. Different types of pheromone traps such as sleeve type trap, delta and sticky traps are also manufactured and sold by different firms.

e. In addition to the above many new pheromones of field and storage pests are being manufactured by commercial firms and will be available to farmers soon.

11.6.7 Biotechnological Approaches

Biotechnology has provided new avenues for management of insect pests and it holds great potential to be included in IPM system.

a. The low toxicity of proteinase inhibitors and Bt alpha-endotoxin as compared to conventional insecticides would reduce the selection pressure and may slow down the development of resistance.

b. Since all plant parts including growing points would remain covered with toxins, dependence on weather for efficacy of the sprays would be eliminated.

c. Since toxins will always be there, there will be no need of continuous monitoring of pests.

d. Transgenic plant would also provide protection to those plant parts which are difficult to be treated with pesticides.

e. Thus, transgenics may prove useful for controlling bollworms and borers which are difficult to control by means of insecticides.

f. The cost of application in the form of equipment and labour will be nil or negative.

g. The development cost is only fraction of the cost of development of conventional pesticides.
h. There would be no problem of contamination in the form of drift and groundwater contamination.
i. Transgenic plants will have inbuilt resistance to various insects replacing some of the current pesticide usage with protection which is intrinsically biodegradable, thus reducing the use of chemical insecticides and minimizing the problem of environmental pollution.

11.6.8 Regulatory Practices

In this process regulatory rules framed by government are brought into force under which seeds and infested plant materials are not allowed to enter the country or from one part to other parts of the country. These are known as quarantine methods and are of two types, i.e., domestic and foreign quarantine.

11.6.9 Pest Surveillance and Monitoring

Pest surveillance is the watch kept on a pest for decision-making.

Objectives

The objectives of the pest and disease surveillance programme in Agricultural and Horticultural crops are to:

a. Detect species of pest or pathogen present.
b. Assess levels of population / damage / infection.
c. Study the influence of weather and seasonal parameters on pests and diseases.
d. Know new species of pests and diseases.
e. Monitor the behaviour of pests under changing cropping pattern / new cultivars.
f. Find out natural enemy population.
g. Watch the behaviour of pests under changing cropping pattern/new cultivars.
h. Assess resistance/susceptibility/break-down of resistance in crops to pests and diseases.
i. Monitor build-up of resistance in pests and pathogens to pesticides.
j. Mark endemic areas/pest calendar.
k. Launch timely plant protection measures on need-basis.
l. Reduce cost of cultivation.

m. Avoid contamination to eco-system.
n. To forewarn farmers.

Methodology for Surveillance

- The surveillance programme encompasses (i) fixed plot and (ii) roving survey. The fixed plot and roving surveys involve the *in-situ* assessment of pests and diseases in the standing crop.
- The supporting methodology includes observations on the activity of pests by setting up light traps, pheromone traps, spore trap, etc. The weather parameters of the respective areas are also recorded to study their influence on the pests and diseases.

11.6.10 Chemical Practices

Use of chemical pesticides is the last resort when all other methods fail to keep the pest population below economic loss. Although there is a great advancement in pest management research, pesticides would continue to play an important role in crop protection in view of complexity of pest problems. Therefore, use of pesticides should be need-based, judicious, based on pest surveillance to minimise not only the cost involved, but also to reduce associated problems. While going for chemical control, we must understand thoroughly what to spray, when to spray, where to spray and how to spray, keeping in mind the following points:

- Pest defender ratio must be observed.
- Relatively safer pesticides should be selected, e.g., neem-based and bio-pesticides.
- If pest is present in strips or isolated patches, whole field should not be sprayed.
- Relevance of IPM practices are more important in vegetable and fruit crops because of their unique mode of consumption by human being.
- CIB and RC approved pesticides should be used against target pest for recommended crops.
- Farmers should not use another pesticide until waiting periods of first pesticides completes and crop should be harvested after completion of waiting period.
- We have to be more careful and cautious in applying pest control practices in field crops.
- When properly used they provide an efficient, fast, reliable and cost-effective means of pest control.

- The draw backs or risks in their use include development of resistance, destruction of natural enemies, poisoning of man and animals, environmental pollution and increasing costs.
- Hence, it is highly needed to switch over to newer insecticide molecules to provide an efficient, fast and reliable means of pest management.

Criteria for newer insecticide molecules

The newer insecticide molecules should have the following criteria for their environmental safety with effective control:

- Safer to natural enemies
- Low mammalian toxicity
- Effective management with low dose
- Broad spectrum
- No / Low residues
- No phytotoxic effect
- Suitable formulation
- Suitable method of application

References

Abrol DP and Shankar U (2012). *Integrated Pest Management: Principles and Practice*. CAB International.

Dhawan AK, Singh B, Bhullar MB and Arora R (2017). *Integrated Pest Management*. Scientific Publishers, India.

Jacobson M and Crosby DG (1971). *Naturally Occurring Insecticides*. Marcel Dekker, New York, USA.

Jayraj S (1995). *Integrated Pest Management for Sustainable Development*. Dr T.V. Ramakrisna Ayyar Birth Centenary Lecture, Entomology Research Institute, Loyola College, Madras, India.

Jayraj S, Ananthakrishnan TN and Veeresh GK (1994). *Biological Pest Control in India: Progress and Prospective*. RGICS project No. 2, Rajiv Gandhi Institute for Contemporary Studies, New Delhi, India.

e-resources

e-Learning portal on Agricultural education, *https://ecourses.icar.gov.in*

What is integrated pest management, *https://www.beyondpesticides.org*

Integrated pest management, *https://em.m.wikipedia.org*

Components of IPM / Directorate of plant protection, Quarantine and Storage, GOI, *http://ppqs.govt.in/divisions*

Outcomes Assessment

Part A

Answer the following questions true or false.

1. Pheromones are chemical substances released by fungi. True/False
2. Predators are free living organisms which prey upon other organisms for their food. True/False
3. Biocontrol is the use of living organisms to control pests. True/False
4. Chemical spray should be adopted as first resort. True/False
5. Insecticide molecules should be safer to natural enemies. True/False

Part B

Answer the following questions.

1. Differentiate between parasitoid and predator.
2. Define bio-pesticide.
3. What is NSKE?
4. Neem Gold contains —% of azadirachtin.
5. Biocontrol practices are grouped into 3/4/6.

Part C

Write a brief note on each of the following.

1. Bio-pesticides.
2. Criteria for newer insecticide molecules.
3. Pheromones in pest management.
4. Biotechnological approaches in pest management.
5. Chemical practices for pest control.

12

Utilization of Resources Constraints in Existing Systems

12.1 Introduction

Performance of an orchard mostly depends on the availability and quality of land and water resources and their management in fruit production. Before taking a decision on all the other possible alternatives under consideration, it is important to undertake feasibility study of the available resources to the orchard management. Other conditions in the growing environment, such as climate and weather conditions, disease and pest management and other factors affecting crop growth also affect production. With ever increasing population in India, today the major problem confronted in fruit cultivation is to improve or at least sustain the production to meet ever increasing demand for fresh or processed fruits. Fruit production can vary widely in response to management of available resources. India cannot afford to mismanage its natural resources (Sharma, 2012). Therefore, judicious utilization of available resources should be done. The following resource constraints are required to be judiciously handled in the management of fruit orchards:

1. Land and Soil management
2. Irrigation water management
3. Solar radiation management

12.2 Land

Land is a vital component of sustainability of any orchard system. Fruit crop must be suited to a particular land and climate in order to achieve a desired success. Sustainable land management is a necessary building block for sustainable agricultural development. Sustainable land management combines technologies, policies, and activities aimed at integrating socioeconomic principles with environmental concerns, so as to simultaneously:

1. Maintain and enhance production (productivity)
2. Reduce the level of production risk, and enhance soil capacity to buffer against degradation processes (stability/resilience)

3. Protect the potential of natural resources and prevent degradation of soil and water quality (protection)
4. Be economically viable (viability)
5. Be socially acceptable, and assure access to the benefits from improved land management (acceptability/equity).

Land use pattern for a particular system depend upon its geographical location such as tropical, sub-tropical, temperate, arid zone, topography, slope etc. Available land for farming should be managed judiciously to get maximum return. If the land is located near to town or city, high density planting or hexagonal planting system should be followed, provided the land must be much fertile. Whereas in sloppy land, farmer should follow terrace planting system. Terrace system of planting in hilly areas helps in minimizing water and nutrient loss, as well as prevent soil erosion. In orchard, a lot of vacant space is available which are wasted. So, in order to enhance the land use efficiency, farmer should go for intercropping. By doing multiple cropping system, farmer not only gets extra income, soil fertility is also improved (by growing leguminous crop). Due to intercropping, weed growth is suppressed as well as soil compactness is minimized. If leguminous crops are intercropped, ample of nitrogen is supplemented to soil, which ultimately reduces the nitrogen fertilizer requirement. Different soil cultural practices such as mulching, intercropping, cover cropping, clean cultivation, sod culture etc. can keep the orchard soil healthy.

Soil erosion is a major problem in orchards, particularly more in sprinkler and flood irrigation systems and in hilly areas. When soil erosion occurs due to water, it leads to transport of soil particles into downstream waterways. These sediments may carry nutrients and unwanted pesticides that adhere to them. There are several management practices that can be employed to control erosion. Improvement of the physical qualities of soil through orchard floor management is an attractive option because it often results in improved yields, better water use efficiency, and reduced runoff. In addition, contour cultivation and terracing in hills, conservation bench terraces can be followed.

Surface of the land become compact due to rain drop, which is responsible for poor vegetative growth. Excessive animal or mechanical traffic, usually with vegetative removal, also compacts the soil surface. Some degree of surface compaction accompanies almost all seasonal crop production. Mechanical disturbance to loosen compacted soil surfaces is also effective. Cultivators use a wide array of indigenous cultivation techniques that loosen the physical hardness of the surface soil to assure timely establishment of their planted seed. However, if bare soil is not protected from the direct impact of raindrops, infiltration rate will quickly decrease during the rain event, and runoff, with its erosion potential, will increase. Infiltration of rainfall into the soil is enhanced if soil is protected from direct raindrop impact. Undecomposed plant residues on the soil surface or complete live plant cover increases infiltration relative to that occurring in bare soil. Earthworm or insect burrows and root channels that open to the surface are effective conduits for infiltration. Higher diurnal soil temperature fluctuations of unshaded soil and physical disturbance usually reduce earthworm and insect activity. Such soil alterations engendered by cultivation practices represent redirection of soil use and favour cultivated food crops over natural vegetation. The cover crops are also effective in correcting soil physical problems such as compaction, which can occur from equipment driving over the previous orchard.

12.3 Soil

Soil is superficial earth crust which functions as store house or reservoir of water and nutrients for plant growth. Soil is a natural laboratory carrying out several biochemical processes to support plant life. Soil microorganism, soil temperature, soil moisture, soil aeration and nutrient levels in the soil are the most important factors in making any soil fertile. The finite quantities of essential elements contained in soil determine its sustainable use for food production. Various soils differ substantially in their ability to provide essential elements. Methods of managing soil to facilitate production of food crops differ depending upon the type of crop grown and the characteristics of the soil. High rates of food production can be achieved only if the concentrations of essential elements in the soil are enhanced. When dedicated to food crop production, soil properties are altered by management practices to favour the requirements of the crop plant and discourage the growth of other vegetation. No soil can sustain the constant depletion of critical elements contained in the plant parts used as food products and transported to another location for consumption. This elemental depletion must be compensated for use to be sustainable (Buol 1995).

Irrigation is one of the oldest soil-management technologies. When irrigation is not conducted correctly, salinization results. In saline soils, proper land drainage (removal of accumulating salts via a drainage system), disposal of drainage effluent, checking drain clogging and irrigation reduces the soil salinity problem. Planting of tolerant tree/crop/grass species is an alternative approach to tackle saline soil.

Soil air and soil moisture are interdependent on each other. A good soil should have 5% organic matter, 45% mineral matter, 25% air and 25% water. The pore spaces in a soil are normally filled with air or water. As the amount of water increases, the amount of air must therefore decrease and vice-versa. The pores of a well-drained soil have certain physical characteristics that, after a period of heavy rainfall, enable water to rapidly drain away and allow air to return to its original percentage. So, proper care should be taken during irrigations so as to maintain proper balance between soil air and soil moisture. Frequent irrigation leads to blocking the pore spaces and reduces soil aeration. So, soil should be stirred after every 2-3 irrigations depending upon the nature of the soil. Soil cultivation or soil culture is another important factor; all horticulturists should have a thorough knowledge of soil culture. Soil culture should mainly depend upon the rainfall of the area. In high rainfall areas, it is better to cultivate soil after rains which help in conserving the soil moisture. Whereas, in low rainfall areas the soil surface should be stirred lightly before rain to prevent the runoff and make the water enter into the soil.

12.3.1 Soil Fertility

Fruit trees will do well only if the soil is deep, well-drained and without hard pan underneath. The best soil is a well-drained loam with a minimum of 3 to 4 feet deep. Good drainage, however, should take preference over depth. Soil fertility should be medium to low. Overly fertile soil can lead to excessive tree growth at the expense of fruit production. Adding fertilizer to increase tree vigour is easier than trying to reduce vigour.

Soil nutrients are mostly lost by crop removal, erosion, leaching, volatilization etc. The loss of nutrients should be replenished by adding of fertilizers, manures, green manures, green leaf manure, incorporating plant residues in the soil etc. Organic matter in the soil plays very important role in maintaining good physical condition and thereby helps in maintaining soil fertility. Soil organic matter contributes to stable soil aggregates by binding soil particles together of various shape and sizes. It greatly improves the soil water holding capacity, decreases the plasticity and cohesion. Soil organic matters strongly influence the population of soil microorganism.

Soil fertility can often be corrected by applying fertilizer or by increasing the level of organic matter in the soil. Fertilizer requirements for tree fruits can be determined by nutrient analysis (fruit, leaf, and soil analysis) and observation of tree performance. Leaf and soil analysis have been the most common means for determining nutrient requirements. Soil reaction or pH is another important factor. Most of the plants prefer nearly neutral pH only. The desired pH range for tree fruit is 5.5 to 6.6 and when the pH gets below or above these levels nutrient disorders can occur. In general, apply limestone (calcite or dolomite) every third year at 5 metric ton/ha (2 tons/acre) until the pH is raised above 5.5 and thereafter is often enough to keep pH within the desired range of 5.5-

6.6. For the purpose of raising and maintaining soil pH use calcitic limestone whenever possible and use dolomitic limestone only if both soil pH and soil Mg need to be increased. When limestone is worked into the soil, enough can be applied to raise the pH to within the desired range. A maximum of 6 metric ton/ha can only be applied as a surface application. When limestone is not applied until after trees are planted in an acidic soil, then the combined effects of manganese/aluminium toxicity, low availability of phosphorus/magnesium, and low rates of nitrification, acting over a period of 5 or more years that will impact negatively on tree growth to such an extent that the trees may never recover. In addition to these, liming has direct beneficial effects on availability of nutrients and prevents toxicity. There are also indirect beneficial effects on the soil like better soil structure, improved drainage and better aeration. One should avoid the too much lime, because availability of boron and zinc can be depressed at a high pH. In heavy rainfall areas, soils tend to become more acid so suitable amendments should be added to the soil from time to time. In many cases, acidic infertile subsoil conditions limit downward root growth. This condition, known as aluminium toxicity, is not easily corrected unless calcium-containing materials such as ground limestone or gypsum are physically incorporated. Liming materials applied at the surface may take several years to neutralize subsoil acidity. Fortunately, many of the more acid soils also have low cation exchange capacity and the downward migration of calcium appears to occur more rapidly in such soils.

Alkali soils can also be managed by adding different soil amendments such as pyrite, gypsum, manures or planting tolerant or resistant crops. Application of balanced nutrition, proper water management practices and right agronomic practices helps in minimizing soil alkaline content.

12.4 Water Resource Management

Water is an important natural resource and is the very basis of our life. A large quantity of water is used for irrigation and there is an urgent need for proper water management in irrigation sector. Over-irrigation through canals has led to water logging in western Uttar Pradesh, Punjab, Haryana and Hirakud command area. Seepage along the canals can be checked by lining them. The overdraft by tube-wells has resulted in lowering of water table in a number of villages in Haryana, Punjab and western Uttar Pradesh. In arid areas, wherever water has been brought for irrigation, saline and alkaline tracts have emerged, rendering the soil infertile. Wasteful use of water should be checked. Sprinkler irrigation and drip irrigation can play a crucial role in conserving scarce water resources in dry areas. Drip irrigation and sprinkles can save anywhere between 30 to 60 % of water.

12.4.1 Principles of Water Harvesting and Conservation

In deciding which techniques to use to make more efficient use of the available water, it is important to consider how crops receive or lose water. Crops receive water through rainfall, irrigation and stored soil water. They lose it through run off, evaporation and drainage. Some key principles on effective water management are:

a. **Use rainwater effectively:** In many climates, rainfall is distributed unevenly in intense downpours that cannot be readily used by a crop. Storage techniques (such as external catchments or roof top collection) increase the availability of water in the drier seasons. They also harvest water from a wider area making more water available to the crop. Measures can also be taken to avoid the rain water running off the surface during intense rainfall.

b. **Make effective use of soil water reserves:** The soil stores water from rainfall providing a reserve that is available to the crop. How much water is available depends on the soil type and the rooting system of the crop. Sandy soils hold much less water than clay or silt soils, so crops will require watering more often. Deeper rooting crops, such as grasses or cereals will exploit soil water reserves more effectively than shallower rooting crops such as vegetable crops and therefore can be grown in drier periods. Good cultivation practices (e.g. not ploughing too deep or when the soil is wet) that result in a soft, friable soil will also promote deep rooting and efficient use of soil water reserves.

c. **Take measures to avoid run off:** Run off is where water is not absorbed by the soil but runs across the surface away from where the crop can use it. Structures such as contour schemes, terracing, pits and bunds can reduce run-off. Run off is more likely to occur on silty or clay soils where the surface has been subjected to intense rainfall then baked in the sun to form a crust or cap. Adding mulch to break up the intensity of rainfall, or adding manure, compost or incorporating green manure residues will reduce the tendency of the soil to form a crust.

d. **Avoid wasting water through evaporation:** Water that evaporates directly from bare soil is wasteful as it is not being used for productive plant growth. It is desirable to maintain full ground cover for as much of the time as practically possible. Applying mulch to the soil will also reduce evaporation considerably. Use of drip irrigation and irrigating in the evening will also reduce the amount of water lost through evaporation.

e. **Reduce water losses through drainage:** When water drains out of the soil, not only is it wasted but essential mobile nutrients such as nitrogen are also washed out. This is more of a problem on light sandy soils. Adding organic matter in the form of compost, manures or plant residues will eventually increase the amount of water a soil can retain, but this will only have an effect if it is added over a longer period of years. Most drainage occurs during the heavy rains, especially if the soil is left bare. Growing a cash or cover crop during this period reduces these losses, as the roots lift water and nutrients back from deeper to shallower soil profiles.

12.4.2 Planning Irrigation

Irrigation is one way of supplementing water from rainfall and soil reserves, but can waste large amounts of water if not used carefully. A key way of making the most of the water supply is to only irrigate when necessary. Many people irrigate on a regular basis whether the crop needs it or not. If water is scarce, irrigation should be restricted to the most critical periods such as germination and fruit set. Fruit drop as well as the size and quality of fruit crops, i.e., mango fruit seems to be influenced by irrigation at certain times. Irrigation during the developmental stage of the fruit is essential to prevent fruit drop and to promote the development of young fruit. Additional irrigation from fruit set to ripening results in a considerable improvement in both fruit size and quality. Drip irrigation makes much better use of water than overhead systems as it is targeted to the roots rather than sprayed up into the air.

Drip irrigation: It is the process of slow application of water in the form of discrete, continuous drops through mechanical devices called emitters or applicators located at selected points along the water delivery line. The key advantage of a drip irrigation system is that water is targeted directly to the root zone; so, applications can be closely controlled. This considerably reduces the amount of water lost through evaporation and runoff as found in conventional irrigation systems. This system saves water around 50-60% compared to traditional irrigation methods. This means, a farmer can irrigate more crop area per unit of water used. So, water use efficiency is increased.

Sprinkler Irrigation: Overhead or sprinkler irrigation refers to the application of water to crops in the form of spray from above the crop like rain. Water under pressure is carried and sprayed into the air above the crop through a system of pipes laid on the ground. In this method, water is used economically, so water use efficiency is high. It can be adopted even on undulating topography. It is very suitable to the area where water is scarce and costly. Soil water can easily maintain at favourable tension for optimum growth and yield. Application of fertilizer and pesticide can be possible. Uniform application of water is possible.

12.4.3 Farming Methods

There are various farming methods through which rain water as well as irrigation water can be conserved and used efficiently. These methods are described below:

***Earth basins*:** Earth basins are designed to collect and hold rainfall and are easy to construct by hand. They are square or diamond shaped basins with earth ridges on all sides. Runoff water is channelled to the lowest point and stored in an infiltration pit. The lowest point of the basin might be located in one of the corners (on sloping land) or in the middle (on flat land). Earth basins are usually used for fruit crops and the seedling is planted in or on the side of the infiltration pit. The size of the basin depends on local rainfall and the water requirements of the trees. They are larger on flat land and smaller on sloping land. They are usually l-2 m long, though sometimes basins of up to 30 m are constructed. Earth basins are suitable in arid and semi-arid areas, with annual rainfall amounts of l50 mm and above. Soils should be deep, preferably at least l.5-2 m to ensure enough water holding capacity. The slope can be from flat up to about 5%.

Fig. 12.1: Earth basins

***Semi-circular bunds*:** Semi-circular bunds are earth bunds formed in U-shapes on a slope. The uppermost tips of the U lie on a contour so that run off is collected in the lowest section of the U. A shallow pit is sometimes also dug in this section to help concentrate moisture. Their size varies from small structures (radius 2 m) used for fruit trees or seedlings to very large structures (radius 30 m) used for rangeland rehabilitation or fodder production. Semi-circular bunds are suitable on gentle slopes (normally below 2%) and uneven terrain in areas with annual rainfall of 350-700 mm. The soils should not be too shallow or saline.

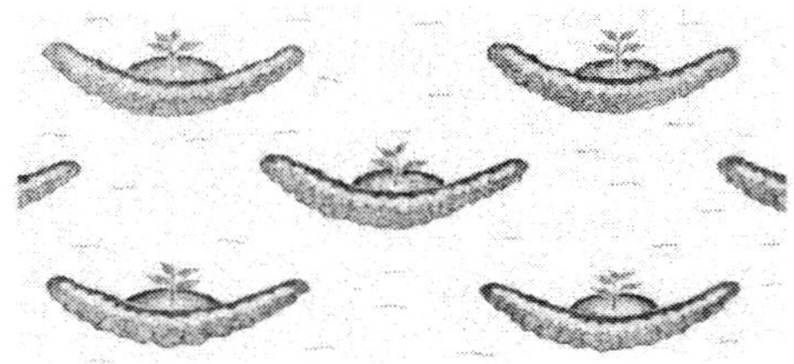

Fig. 12.2: Semi-Circular Bunds

***Mulching*:** Mulching means covering the soil between crop rows or around trees with a layer of loose material such as dry grass, straw, crop residues, leaves, manure or compost. This helps to retain soil moisture by limiting evaporation, suppressing weed growth and enhancing soil structure, reducing runoff, protecting the soil from splash erosion and limiting the formation of crust. In addition, mulching reduces fluctuations in soil temperature which improves conditions for micro-organisms. It is commonly used in areas affected by drought and weed infestation. It is most effective if applied at the start of the rains, as it intercepts and increases water take-up, but it is frequently more practical to mulch towards the end of the rains when grass is available.

12.4.4 Contour Farming

In hilly areas, contour farming is most effective method to conserve the soil moisture and prevent runoff. Contour farming refers to field activities such as ploughing and furrowing that are carried out along contours rather than up and down the slope. They conserve water by reducing surface run off and encouraging infiltration of water into the crop area. A number of water-harvesting techniques are based along contours including: contour ploughing; contour ridges; stone lines; grass strips and terraces. The technique used depends on the steepness of the slope, soil type, conditions, crops grown and other factors such as the availability of labour.

i. *Contour Ploughing*: Ideally, any ploughing on a slope should be carried out along the contours rather than up and down as this reduces run off and soil erosion and increases moisture retention. Contour ploughing can be practiced on any slope with a gradient less than 10%. On steeper slopes it should be combined with other measures such as terracing, bunds or strip cropping. It is not always carried out in practice because the shape and topography of the field may be considered a more important factor in determining the direction of ploughing. It is important to lay out contours properly or they may channel the water and increase run off. It is not suitable for heavier soil types with low infiltration and also not suitable for very low rainfall areas as the catchment area is limited. Structures may be breached by intense rainfall.

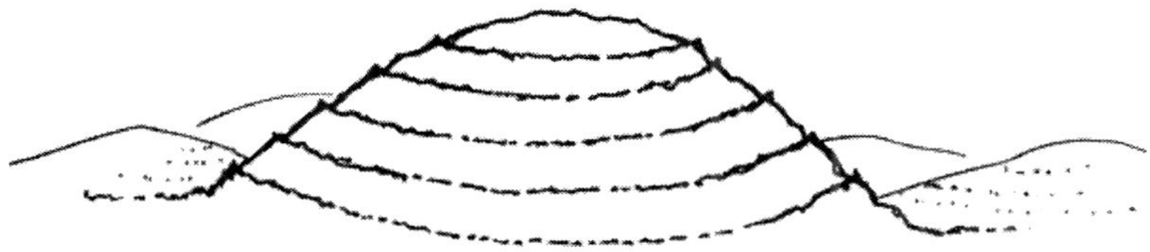

Fig. 12.3: Contour Ploughing

ii. *Contour Furrows*: Contour furrows are small earthen banks that run along a contour. A furrow is dug next to each bank on the upper side of the slope. The distance between the ridges varies between 1-2 m depending on the rainfall and the slope. The aim of contour furrows is to concentrate moisture into the ridge and furrow area where the crops are planted by trapping run off water from the catchment area between them. This also decreases the risk of erosion. Plants with higher water requirements, such as peas or beans, can be planted on the higher side of the furrow whereas cereal crops requiring less water, such as sorghum or millet, can be planted on the ridges. The catchment area between the ridges should be left uncultivated and clear of vegetation so that run off into the ridges is maximized. Under drier conditions, the furrows are spaced further apart to harvest water from a larger catchment area. Contour furrows are suitable for areas with lower rainfall (350 – 700 mm) and gentle slope (0.5 –3%). However, the amount of water harvested is limited, so they are not suitable for very dry areas. Extreme rainfall may cause the ditches to overflow and break. This is more likely to occur on heavier soils with a lower infiltration rate, or on steeper slopes. The risk can be reduced by building higher ridges, although this increases the labour requirement.

Drought sensitive plants on side of furrow Drought resistant plant on ridge

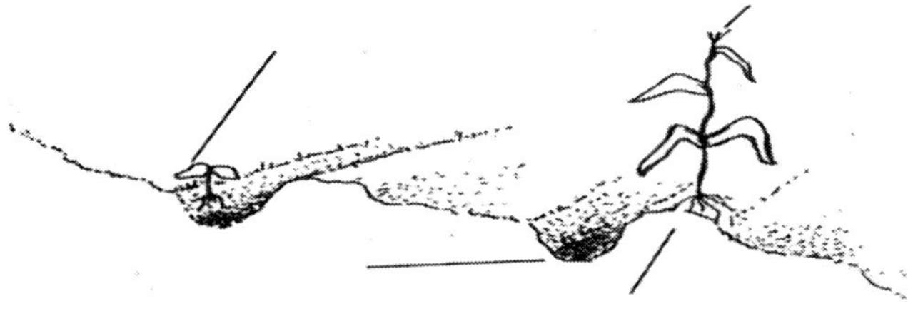

Fig. 12.4: Contour Furrows

iii. *Bench Terraces*: Terraces are made by creating ridges and furrows along contours on a slope. The ridges hold back water and soil runoff and eventually turn the hillside into a number of terraces. These can be stabilized by planting grasses or shrubs on them. Terraces can be used on steeper slopes (up to 40%) than other contour methods, but building them requires very high labour input. Terraces are formed by digging a ditch along a contour and throwing the earth either uphill or downhill to form a ridge. Water is stored behind the ridge. If the earth is thrown uphill the terraces are suitable for steep slopes with gradients of 30-55%. If the earth is thrown downhill there is a higher risk of the ridges breaking due to water pressure. So, this method is not suitable for steep slopes (up to 35% gradient) or areas with intense rainfall.

Fig. 12.5: Bench Terraces

iv. Grass Strips: Strips of grass (up to 1m wide) planted along a contour can reduce soil erosion and runoff. Silt builds up in front of the strip and over time benches are formed. On gentle slopes (1-2 %) the strips should be widely spaced (20-30 m apart), and on steeper slopes narrowly spaced (10-15 m apart). Grass strips are most likely to be used in areas where fodder or mulch is also needed. Many grass varieties can be used, depending on what is locally available, e.g., Vetiver, Napier, Guinea and Guatemala grass. They are not suitable for steep slopes or in very dry areas since grasses will compete with the crop.

v. Stone Lines: Stone lines running along the contour are one of the simplest contour techniques to design and construct. The lines of stones form a semi-permeable barrier that slows the speed of runoff so that spread of water over the field and infiltration is increased, and soil erosion reduced. Stone lines are suitable on gentle slopes (0.5-3%) in areas with annual rainfall of 350- 700 mm.

vi. Retention Ditches: Retention ditches work on a similar principle to contour furrows but on a larger scale. They are large ditches, designed to catch and retain all incoming run-off and hold it until it infiltrates into the ground, increasing the supply of water to crops planted in the ditch and reducing soil erosion. They vary from 0.3-0.6 m deep and 0.5-1 m wide and suitable in shallow slope (<2%). They are usually used on flat land where they may be spaced at 20 m or on gentle slopes where the spacing can be decreased to 10-15 m. Retention ditches are commonly used in semi-arid areas for growing crops that have high water requirements, such as bananas.

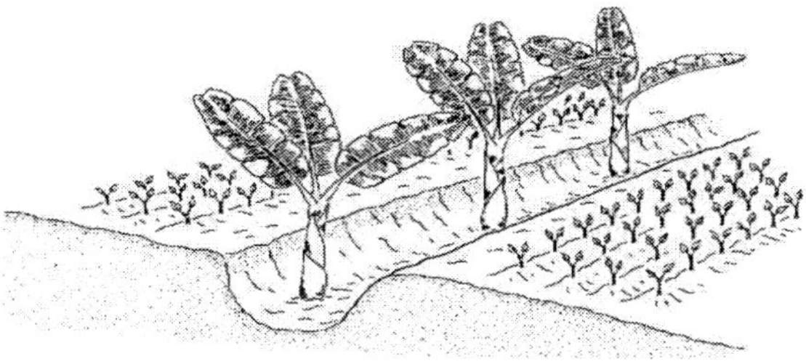

Fig. 12.6: Retention Ditches

12.4.5 Rain Water Harvesting

In many farming areas, readily available water is in short supply. Although the total annual rainfall in an area may be enough to sustain farm needs, it is often distributed very unevenly so that long dry periods are interspersed with periods of intense rainfall. In many cases, a crop is unable to use a high proportion of this water, as much of it is lost through run off or leaching.

Rain water harvesting is one of the most effective methods of water management and water conservation. It is the term used to indicate the collection and storage of rain water used for human, animals and plant needs. It involves collection and storage of rain water at surface or in sub-surface aquifer, before it is lost as surface run off. The augmented resource can be harvested in the time of need. In fruit plantations, rain water can be harvested in the tree basin areas by certain soil working techniques such as preparation of tree basins. A well-prepared tree basin prior to rains, aid in water infiltration and subsequent frequent shallow hoeing not only remove the competitive weeds but also prevent evaporation losses of moisture from the soil by forming dry soil mulch over the sub-soil. For arid and semi-arid regions, soil working techniques like 'Crescent Bund with Open Catchments Pits', 'Trench Systems', 'V-ditch' etc. provide satisfactory in-situ harvest of rain water. These techniques are found to be useful for conserving soil moisture in the root zone of trees for the dry months.

12.4.6 Water Harvesting from External Catchment

Water harvesting from external catchments involves diverting runoff water from an area that is not cropped to the area where crops are grown. Water is stored in a simple reservoir structure/ pond and can be applied to the crops when it is needed. The flow of water from the reservoir into the cropped area can be controlled using tied bunds that can be built up or dismantled as required. The external catchment area should not be cultivated and may include rough grazing areas, roads or homesteads. Ideally the soil should have a low infiltration rate in order to maximize runoff and therefore vegetation should be restricted to a minimum. The stored water will be lost gradually through evaporation and seepage, this can be reduced on silt or clay soils by capping the soil using puddling boards (used to encourage flooding in rice cultivation). Suitable sealants may be used to check/reduce seepage losses in the structure. It may be lined with polyethylene sheet (covered with soil or bricks or round boulders to protect it against UV sun light), silpauline sheet (UV resistant), bitumen, cement and concrete lining, RCC, etc. The cost and longevity of structure will depend upon the lining material used.

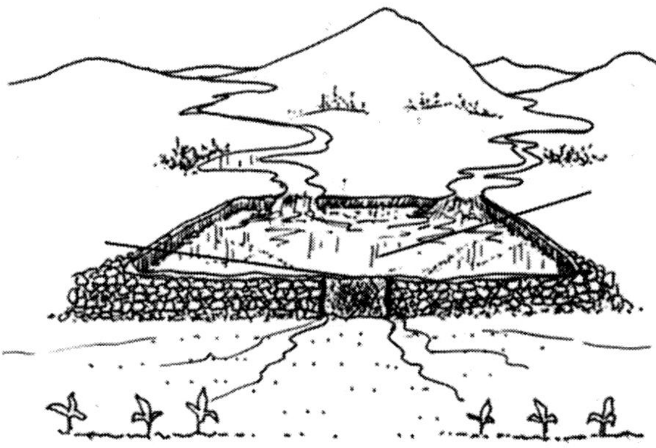

Fig. 12.7: Water Harvesting from External Catchment

12.5 Solar Radiation Management

The productivity of plant greatly depends upon the absorption of light energy by the chlorophyll of the leaf tissue and conversion of that solar energy into biomass via photosynthesis. It has been observed that the fruit yield is linearly related to light interception. Although high light interception is required for increasing the productivity, shade can have an adverse effect on fruit quality, fruit set and flower initiation. Shading may arise from within or between tree sources, including windbreaks. Fruit size, red skin colour, and soluble solids concentration are all influenced by light, i.e., apple required UV light for development of red colour on skin, and oranges and grape required high light intensity during fruit ripening for high sugar and TSS content in fruit. However, some fruit crop needs partial shade for good yield, i.e., pineapple. This seems to be a general phenomenon among perennial tree fruits and has been reported for apples, citrus, red raspberry, kiwifruit, cherries, peaches and grapes. In order to improve the yield, we have to manage solar light judiciously. An unmanaged larger canopy receives less amount of solar radiation as compared to a well-managed tree. For this, farmer has to regulate the tree canopy architecture, so that maximum solar radiation can be penetrated inside the canopy and all the leaves remain photosynthetically active. There are different training systems such as open system, modified central leader, central leader, bush system etc. followed to develop the tree architecture which depends on tree growth habit and climatic condition. In temperate fruit crops, conical tree shapes, Telephone system, Tatura trellis, V system, Spindles, Kniffen system, etc. ensure better light penetration and interception and therefore production of higher quality fruits. Along with training system, planting density also influences the quantity and quality of light interception inside the canopy. By adopting high density planting system along with maintaining dwarf architecture the productivity of the land can be improved.

References

Buol S W (1995). Sustainability of soil use. *Annual Rev. Ecol. Syst.*, **26**: 25-44.

Duveskog D (2003). Soil and Water Conservation with a focus on Water Harvesting. A Study Guide for Farmer Field Schools and Community-based Study Groups. Version 1.1 FARMESA, Harare, Zimbabwe. http://www.infobridge.org/ffsnet/output_ view. asp?outputID=1950.

Outcomes Assessment

Part A

Answer the following questions true or false.

1. Land is not a vital component of sustainability of any orchard system. True/False
2. The fruit yield is linearly related to light interception. True/False
3. Retention ditches work on a similar principle to contour furrows. True/ False
4. Terraces are made by creating ridges and furrows across contours on a slope. True/False
5. Soil air and soil moisture are independent of each other. True/False

Part B

Answer the following questions.

1. What is contour farming?
2. Terraces are made by creating ridges and furrows along/across contours on a slope.
3. Define mulching.

4. What is the difference between drip and sprinkler irrigation?
5. Water is an important — resource.

Part C

Write a brief note on each of the following.

1. Rain water harvesting.
2. Principles of water harvesting and conservation.
3. Solar radiation management.
4. Earth basins.
5. Water harvesting from external catchment.

13

Climate Aberrations and Mitigation Measures of Horticultural Crops

13.1 Introduction

Climate change is emerging out to be the greatest challenge to humanity. Climate change may be a change in the mean of the various climatic parameters such as temperature, precipitation, relative humidity, atmospheric gases composition etc. and in properties over a longer period and a larger geographical area. It can also be referred to as any change in climate over time, whether due to natural variability or because of human activity. Climate change is a global challenge threatening sustainability of agriculture and its impact is realized more in the fragile and vulnerable ecosystems. The resource poor small and marginal farmers are more prone to climatic aberrations than other farmers. Predicted population growth over the next 50 years will raise food demand globally higher (Trostle 2008), which is complicated by environmental changes such as climate, biodiversity, water availability, land use, atmospheric and other pollutants, and sea level rise (Misselhorn *et al.* 2012). Climate change has emerged as a serious global environmental issue having impact on all forms of life. It increases the greenhouse gases like carbon dioxide, nitrous oxide, ozone and methane which may cause impact in terms of increased temperature, more demand for water and increase in biotic and abiotic stresses. It has direct impact on agriculture and horticulture. Due to climate change, low production of horticultural crops is featured. Due to severe cold wave, horticultural crops suffer a yield loss of 10-100% depending upon crop and variety (Hazarika 2013). Mitigation strategies aim to reduce the impact of changes to climate and the associated risks from anthropogenic (human) processes. Adaptive capacity is defined as "the ability of a system to adjust to climate change (including climate variability and extremes), to moderate potential damages, to take advantage of opportunities, or to cope with the consequences" (IPCC 2001, p. 982). Rise in temperature is likely to be around 2-2.5°C by end of this century along with uncertainties in rainfall and its distribution. Effects of climate changes, even smooth trends, will not be uniform in space or time. Occurrence of heat and cold waves, floods and droughts are not uncommon. Effects of extreme weather events on crops are very severe and may lead to considerable loss to agricultural production to challenge food security. Human interferences such as industrialization, deforestation, faulty agricultural practices, heavy use of fertilizers have increased atmospheric concentrations of greenhouse gases (GHG) such as carbon dioxide, methane and nitrous oxides. Climate change has become a common phenomenon now-a-days as experienced due to increasing CO_2 level, temperature, erratic rainfall, frequent drought and other extreme weather events. In India, as per prediction crop yield will be reduced by 4.5 to 9 per cent, which is roughly up to 1.5 % of GDP per year (Venkateswarlu *et al.* 2013) with medium-term climate change (2010-2039). Climate change impact will be more negative on rainfed agriculture which constitutes nearly 58 % of net cultivated area. The adaptation strategies would initially benefit from the technologies developed so far in this fragile ecosystem to combat drought and desertification, especially the improvements made in the traditional wisdom for water resources conservation and use, agro-forestry systems and mixed cropping, etc. Efforts would then be required to enhance crop resilience and adaptation to climate change for sustainable production, food and nutrition security. This may be achieved by developing cultivars of arid crops that are tolerant to high temperature during flowering time, extra early cultivars for shortened moisture availability period and cultivars with terminal drought tolerance. Climate risk resilience can be built in horticulture production systems by promoting crop diversification from rice to low water requiring horticultural crops like vegetables and fruits; cultivation of high yielding, short duration, drought-, pest and disease tolerant crop cultivars / hybrids; complementary agronomic practices; and building ability to withstand biotic and abiotic stress through a large scale program of demonstrations, and training activities as well as investments in farm implements and resource saving technologies like micro irrigation systems.

13.2 Impact of Climate Change on Horticulture

Global warming and climate change are the greatest concern of mankind in 21st century. The established commercial cultivars of fruits, vegetables and flowers will perform poorly in an irregular manner due to aberration of climate. Melting of ice cap in the Himalayan regions will reduce chilling effect required for the flowering of many of the temperate horticultural crops like Apple, Saffron, Rhododendron, Orchid, etc. Commercial production of horticultural plants particularly grown under open field conditions will be severely affected. Inter-Governmental Panel on Climate Change has predicted that by the end of this century, global earth temperature is likely to increase by 1.8 to 4.0°C. This will cause frequent drought, heat waves, erratic rainfall and the incidence of pests and diseases would be significantly increased. This will cause instability in food production and will threaten farmers' food security. In India, as per prediction crop yield will be reduced by 4.5 – 9.0 %, which is roughly up to 1.5 % of GDP per year (Venkateswarlu *et al.* 2013) with medium-term climate change (2010-2039). Climate change impact will be more negative on rainfed agriculture which constitutes nearly 58 % of net cultivated area.

Due to high temperature physiological disorder of horticultural crops will be more pronounced, e.g., spongy tissue of mango, fruit cracking of litchi, flower and fruit abscission in solanaceous fruit vegetables, etc. Air pollution also significantly decreases the yield of several horticultural crops and increases the intensity of certain physiological disorder like black tip of mango. Hence there is a need to protect these valuable crops for sustainability against the climate change scenario. The most effective way is to adopt conservation agriculture, using renewable energy, forest and water conservation, reforestation etc. To sustain the productivity, modification of present horticultural practices and greater use of greenhouse technology are some of the solutions to minimize the effect of climate change.

The yield reduction is likely to be brought about by many factors including pests, weeds and diseases, loss of biodiversity, rise in sea level, saline water intrusion in coastal belts, poor quality of irrigation water, decline in soil fertility, and irregularities in onset of monsoon, heat wave, cold wave, drought, flood and cyclone. The global warming has caused loss of vigour, fruit bearing ability, reduction in size of fruits, less juice content, low colour, reduced shelf life and increasing attack of pests resulting in low production and quality of apples (Hazarika 2013). According to him, in mango, unusual or very early flowering, leaf scorching, twig dying are common symptoms of heat stroke. In guava, there is severe increase in pests and diseases due to climate change. Fruit fly in guava is becoming alarming due to hot and humid conditions. The crops like peach, plum, which require low chilling temperature are also showing sign of decline in productivity. Abnormal high temperature during winter causes poor flowering, irregularity in flowering duration, pattern of flowering and poor yield in pear due to non-availability of sufficient chilling hours during winter months. The horticulturists will have to play a significant role in the climate change scenario and proper strategies have to be envisaged for saving horticultural crops. Developing new cultivars tolerant to high temperature and producing good yield under stress conditions should be the main strategies to meet these challenges (Hazarika 2013). The climate change will probably lead to a decrease in crop productivity, but with important regional differences (McCarty *et al.* 2001). In tropical and sub-tropical regions like in India where the crops are already near the limit of their temperature tolerance, even a slight increase in temperature will result in drastic fall in crop productivity. However, crop productivity is expected to rise slightly in mid to high latitudes for mean temperature increases of up to 3°C coupled with enhanced CO_2 concentration, food productivity in these two areas is expected to increase with rise in temperature up to 3°C and fall with further rise in temperature. Small changes in temperature and rainfall will have significant impact on quality of food grains, vegetables, tea, coffee and medicinal plants with resultant implications in domestic and external trade.

Table13.1: Potential impacts of climate change on different sectors of agriculture and horticulture

Sector	Impact
Crop	Increase in ambient CO_2 concentration is beneficial since it leads to increased photosynthesis in several crops and decreased evaporative losses. Increase in extreme weather events such as floods, droughts, cyclones and heat waves will adversely affect agricultural and horticultural productivity. Quality of fruits, vegetables, tea, coffee, aromatic, and medicinal plants may be affected. Agricultural and horticultural biodiversity is also threatened due to the decrease in rainfall and increase in temperature, sea level rise, and increased frequency and severity of droughts, cyclones and floods.
Water	Demand for irrigation water would increase with rise in temperature and evapo-transpiration rate. The melting of glaciers in the Himalayas will increase water availability in the Ganges, Brahmaputra and their tributaries in the short-run, but in the long run, the availability of water will decrease considerably. This additional water in the wet season, on the other hand, may lead to increase in frequency and duration of floods.
Soil	Organic matter content, which is already quite low in Indian soils, would become still lower. Quality of soil organic matter may be affected. The residues of crops under the elevated CO_2 concentrations will have higher C:N ratio, and this may reduce their rate of decomposition and nutrient supply. Rise in soil temperature will increase N mineralization, but its availability may decrease due to increased gaseous losses through processes such as volatilization and denitrification.
Livestock	Climate change will affect fodder production and nutritional security of livestock. Increased temperature would enhance lignification of plant tissues, reducing the digestibility. Increased water scarcity would also decrease production of feed and fodder. Major impacts on vector-borne diseases will be through expansion of vector populations in the cooler areas. Climate change is likely to aggravate the heat stress in dairy animals, adversely affecting their reproductive performance.
Fishery	Increasing temperature of sea and river water is likely to affect breeding, migration and harvests of fishes. Impacts of increased temperature and tropical cyclonic activity would affect the capture, production and marketing costs of the marine fish.

Source: Aggarwal *et al.* (2009)

13.3 High Temperature Effect

According to studies carried out by the Intergovernmental Panel on Climate Change (IPCC), average air temperatures will increase between 1.4 and 5.8°C by the end of this century, based upon modelling techniques that incorporated data from ocean and atmospheric behaviour (Anon 2001). Higher temperatures can increase the capacity of air to absorb water vapour and consequently generate a higher requirement for water. Higher evapotranspiration indices could lower or deplete the water reservoir in soils, creating water stress in plants during dry seasons. For example, water stress is of great concern in fruit production, because fruit trees are not irrigated in many areas around the world. Water stress not only reduces crop productivity but also tends to hasten fruit ripening (Henson 2008). Exposure to higher temperatures can cause morphological, anatomical, physiological, and ultimately biochemical changes in plant tissues and as a result, can affect growth and development of different plant organs. These events can cause drastic reductions in yield. However, by understanding plant tissues physiological responses to high temperatures, mechanisms of heat tolerances and possible strategies to improve yield, it is possible to predict reactions that will take place in the different steps of fruit and vegetable crops production, harvest and postharvest (Kays 1997). Besides increase in temperature and its associated effects, climate changes are also a consequence of alterations in the composition of gaseous constituents in the atmosphere. Due to rise in temperature, photoperiods may not show much variation. As a result, photosensitive crop will mature faster. The winter regime and chilling duration will reduce in temperate regions affecting the temperate crops. The requirement of yearly irrigation will increase and heat unit requirement will be achieved in much lesser time. Higher temperature causes floral abortion, flower and fruit drop, reduces tuber initiation in potato and reduces quality in tomatoes and pollination in many crops. In summer the leaves, new shoots, branches, flowers and fruits are exposed to sun and high temperature. High temperature is generally more harmful when accompanied by low humidity. This causes excessive transpiration and wilting of leaves and twigs. The ability of plants to withstand such condition varies greatly with species; e.g. banana is more sensitive than guava plants. Young trees are affected more than old plants.

The adverse effect of climate change due to high temperature can be avoided by planting of thick and tall growing wind breaks at the time of orchard establishment; avoiding severe pruning of trees during summer season; white washing of the main stem, small branches and young plant shoots at the beginning of summer season and application of irrigation and mulching of tree basins to have more humidity in the orchard.

13.4 Low Temperature Effect

Low temperatures (Freezing) occurring either during the rest period of fruit plants or after wards, cause damage to roots, stems and buds. At lower temperatures, water is withdrawn from protoplasm and ice is formed in the intercellular spaces. If this process proceeds beyond critical limits the protoplasm of the cells disintegrates and dies from loss of water. Frost causes damage to the newly opened blossoms of fruit trees, to young growth and newly set plants.

Frost damage can be mitigated by planting frost tolerant cultivars in the new orchard; regular irrigation in orchards during the frost periods; burning of dry leaves and twigs in orchards and planting of quick growing green manuring crop around young plant.

13.5 CO_2 Effect Due to Climate Change

Carbon dioxide is the major contributor among the greenhouse gases causing global warming. Intergovernmental Panel on Climate change (IPCC, 2007) estimates that the current atmospheric CO_2 concentration has been increased more than 50% with respect to the preindustrial concentration. The expected carbon dioxide concentration in 2100 is estimated to be 100% higher than the one observed at the pre-industrial era (IPCC, 2007 a). Carbon dioxide (CO_2) and ozone (O_3) concentrations in the atmosphere are changing during the last decade and are affecting many aspects of fruit and vegetable crops production around the globe (Lloyd and Farquhar 2008). Increase in atmospheric CO_2 is reported not only to improve the yield but also to alter the quality of the produce. The quality (carotene, starch and glucose content) starchy fruits increased in elevated CO_2 conditions. Under high CO_2 level, the water use efficiency of crops is reported to increase. In coconut, areacanut and cocoa are reported to benefit due to CO_2 concentrations. Studies indicated that the elevated CO_2 (550 and 700 ppm) increased growth of coconut seedling by 20 to 30 %.

13.6 Technological Options for Climate Change Adaptation

1. *Development of drought and heat tolerant cultivars:* Developing climate resilient cultivars with early maturity tendency and potential to produce crop under dry periods and having the ability to make best use of high moisture availability during good rainfall years, thus producing higher yield besides being tolerant to high temperature at flowering and fruiting. Amalgamation of molecular biology, genetics, genomics and plant physiology with breeding is essential to develop the climate resilient genotypes. Such strategies will have combined benefits for climate change mitigation and adaptation.
2. *Crop diversification and intercropping:* In the present scenario of climate change mono- cropping is risky in the low rainfall zone resulting in

lower yield. In such areas intercropping is a feasible option to minimize the risk of crop production. It ensures reasonable returns at least from the intercrop and also improves soil fertility with a legume intercrop. Several underutilized horticultural crops which are important in human nutrition and not addressed have huge potential for crop diversification.

3. *Integrated farming system:* Integrated farming system approach that involves the combination of crops, livestock, trees, fruit crops, vegetables, etc. depending upon the farm resources should be developed. Horticulture-based farming system is considered effective strategy for improving productivity, creating employment opportunities, improving economic condition and providing nutritional security.
4. Development of early maturing, and high temperature and drought tolerant cultivars.
5. Development and introduction of water use efficient crops and cropping systems.
6. Availability of good quality seed and planting materials at affordable prices and providing effective measures to check the flow of unauthentic seeds to farmers.
7. Development of effective insect and disease management strategies in view of changing climate scenario.
8. Popularization of efficient irrigation methods like drip and sprinkler, water harvesting and cultivation of crops that use water more efficiently.
9. Exploitation of underutilized horticultural crops as additional source of income for providing livelihood security.
10. Sustainable intensification and diversification of crop and horticulture production systems can be achieved by awareness creation, on-farm demonstrations of promising technologies, capacity building and training activities and by encouraging private sector investments in water management and farm mechanization.

13.7 Climate Change Effect on Fruit Crops

Global climate changes are likely to exert pressure on fruit production system and may restrict in accomplishment of future fruit production targets. These changes are natural but its control is in our hand through several mitigation measures which reduce the concentration of gases in atmosphere that are responsible for climate change and fruit crops have a great role in mitigation of these gases through carbon sequestration by photosynthesis. The fruit growers encounter many problems due to sudden climate change in their locality. Production of high-quality fruits and higher productivity of the any crop is possible only under favourable climatic condition prevailing in the locality. The changed climatic parameters affect the crop physiology, biochemistry, floral biology, biotic stresses like pest and disease incidence etc. and ultimately result in the reduction of yield and quality of fruit crops. Mitigation is the most important measures to reduce the devastating effect of climate change. Carbon sequestration has the crucial role to mitigate the effect. Fruit crops are highly potential for carbon sequestration (Rajatiya *et al.* 2018).

13.7.1 Flowering and Fruiting

Flowering and fruiting are the most important events in all fruit crops which is regulated by climatic condition. Changed climatic parameters disturb the flowering pattern; fruit setting and pollination in many fruit crops through reduce pollinator activity, pollen viability. Rains during flowering wash out the pollen from stigma of flower and result in poor or no fruit setting. Mango production loss of 80-90% was reported in Gujarat due to unseasonal rain followed heavy dew attack during flowering season; which reduced fruit setting, increased fruit drop at pea stage and also increased heavy incidence of sooty mould and powdery mildew in mango (Varu *et al.* 2015). The intensity of relative humidity was also found higher resulting in higher infestation of mango hopper and powdery mildew.

13.7.2 Physiological Disorder

Due to high temperature several physiological disorders of fruit crops are more pronounced, e.g., spongy tissue of mango, fruit cracking, black spot in custard apple, flower and fruit abscission, etc. Air pollution also significantly decreased the yield of many fruit crops and increased certain physiological disorders like black tip of mango. The mango fruits harvested from clean cultivation showed 15.55 and 18.33% occurrence of spongy tissue under Paria and Ghadoi conditions respectively because of barren soils under clean cultivation showed the mean soil surface temperature to the extent of 52.40°C (Katrodia and Sheth, 1989). The rate of transpiration and respiration has an influence on fruit temperature, which in turn has an influence on normal fruit ripening processes and leads to development of spongy tissue (Hemnath *et al.* 2008). High temperature and moisture stress also increase sunburn and cracking in apples, apricot and cherries and increase in temperature at maturity will lead to fruit cracking and burning in litchi (Mark and Marin (2016). In custard apple maximum black spot (35.63%) on the skin of fruit reported in the year 2009 may be due to high wind speed recorded in the same year (Varu *et al.* 2010).

13.7.3 Crop Duration

A crop duration or maturity indices are the important physiological processes in fruit crops which directly affects the fruit quality and yield. Production timing of crops will change because of rise in temperature; crops will develop more rapidly and mature earlier (Hatfield and Prueger 2015). In lowland tropical areas, due to high respiration rates at warm temperatures, citrus fruit mature quickly and do not have sufficient time to accumulate high TSS, and acidity declines rapidly so that the soluble solids/acid ratio increases sharply, and the fruit quickly become insipid and dry (Zerki 2011).

13.7.4 Pest and Diseases

Rising atmospheric CO_2 level and climate change may also impact indirectly on crops through effects on pests and diseases. Pests such as aphids (Newman 2004) and weevil larvae (Staley and Johnson 2008) respond positively to elevated CO_2 and change climatic parameters also increased threat of new attack. The hot and dry climate reduces the risk of fungal diseases like anthracnose and powdery mildew in mango because sunlight, low humidity and temperature extremes (below 18°C or greater than 35°C) rapidly inactivate spores (Alfonso and Brent, 2014).

13.8 Mitigation of Climate Change Effect in Fruit Crops

Climate mitigation means steps to reduce or stop the emission or enhance the sinks of greenhouse gases which are responsible for climate change with the help of mindful practices, which permanently eliminate or reduce the long-term risk and hazard of climate change to human life and their properties (Rajatiya *et al.* 2018).

The various ways to mitigate climate change through orchard crops as described by Rajatiya *et al.* (2018) are mentioned below:

a) *Improving energy use efficiency* by increasing fuel efficiency in agricultural machinery, by use of wind and solar power or by minimum or no tillage, etc.

b) *Fertilizer, manure and biomass management* through reduced use and production of synthetic fertilizers, escaping leaching and volatilization of N from fertilizers during storage and application, use slow-releasing fertilizers (Pathak *et al.* 2010).

c) *Soil management for increasing soil carbon (%)* by using organic fertilizers, reduced tillage, avoid soil compaction, use of biofertilizer, cover crops, intercropping and other cropping system.

13.9 Adaptation to Climate Change Effect in Fruit Crops

Adaptation is nothing but to include the actions of adjusting practices, processes, and capital in response to the threat of climate change, as well as responses in the decision environment, such as changes in social and institutional structures or altered technical options that can affect the potential or capacity for these actions to be realized. There are several adaptation measures that can be undertaken in the agriculture or horticulture sector to cope up with future climate change (Rajatiya *et al.*, 2018) and they are mentioned below:

1. Adaptation based on development of tolerant or resistant cultivars/varieties/rootstock against climate change and planting different varieties or crop species.
2. Several drought hardy fruit crops like *Capparis decidua, Salvadora oleoides, Cordia dichotoma, Zizyphus nummularia var. rotundifolia, Z. mauritiana* are suitable for cultivation in the area receiving rainfall < 300 mm. Several other fruit such as aonla, pomegranate, bael, date palm, Karonda and tamarind can be grown in the semi-arid areas.
3. Modifying crop management practices, *viz.*, modifying date of planting or date of sowing, adjusting cropping season and off seasonal production and marketing of horticultural crops; using sustainable, customized or liquid fertilizer ; implementing new or improving existing irrigation systems like drip irrigation; improvement in crop residue and weed management and changes in land use management practices; adopting new farm techniques, resource conserving technologies (e.g. bagging of fruits, fertigation, etc.) and by following improved pest and disease management
4. Mulching conserve the soil moisture, improve soil microclimate and soil microbial activity, maintain soil health and also help in controlling weed.
5. Wind breaks or shelter belts which modify the microclimate of orchard as well as soil. Windbreaks also provide shelter for pollinating insects, protect orchard from wind erosion and other natural disaster.

5. Weather forecasting and crop insurance schemes for farmers and use of GIS.
6. Recycling of waste water and solid wastes in agriculture and use water harvesting technologies.

13.10 Effect of Climate Change on Tropical, Subtropical and Arid Fruits

Climate changes have negative impact on tropical regions where the prevailing temperature is already high. Well known traditional fruit growing areas experience risks of abnormal flowering, fruit set and reduced yield with a risk of untimely rains deteriorating fruit quality and giving rise to disease and pest build-up. Cool temperatures also adversely affect germination and pollen tube growth, which is completely inhibited at temperatures below 15°C (Issarakraisila and Considine 1994). In mango most of the cultivars set fruit poorly when exposed to unfavourable environmental conditions of rainfall, humidity, temperature, light, wind, drought, water logging, etc. In the sub-tropics, a sudden rise in day temperatures or low temperatures (less than 10°C) in the night during flowering may increase the number of stenospermocarpic fruit. Self-pollination in mango cv. Dashehari may be one of the causes of clustering, which is further enhanced by the reduced pollinator activity due to low or high temperature. An increase in rainfall during fruit growth and development may delay the number of days taken to maturity and fruits may lose their attractive appearance if exposed to several rains during fruit growth and development periods. Thus, the time taken for attainment of fruit maturity and appearance are likely to be misshapen under the changing climate scenario. Cool temperature during inflorescence development reduces the number of perfect flowers and these flowers produce aborted, deformed and fused ovaries, which does not happen above 17°C. Polyembryonic mango cultivars suffer more from low temperature than monoembryonic (Whiley and Schaffer 1997). Changes in rainfall patterns can adversely affect the quality and appearance of ripe mango fruits. Unseasonal rains encourage pests, which also lower fruit yield. Although mango tolerates a wide range of climates from warm temperate to tropical, anthracnose can become a serious problem for mango cultivation in humid and high rainfall environments (Ploetz 2003). Spongy tissue, a physiological disorder in mango cv. Alphonso, is induced because of build-up of high temperatures within the fruits, leading to tissue breakdown. Banana cultivation may suffer from high temperature, soil moisture stress or flooding or water logging (Diengngan and Hasan, 2014). Due to rise in temperature, crops will develop more rapidly and mature earlier. For example, citrus, grapes, melons etc. will mature earlier by about 15 days (Datta, 2013). Malic acid concentrations were shown to be higher in grape vines growing in cool rather than warm regions and during berry growth, water deficit restricts cell division, reducing berry size (Diengngan and Hasan 2014).

13.11 Adaptation and Mitigation Strategies

Selection of cultivars should be followed on the basis of their adaptability under particular climatic conditions (drought, high temperature) of a particular location. Rootstock should be selected on the basis of tolerance to drought and salinity. It should be low in vigour having strong root system. In tropical climate without pronounced seasonality, it is possible to artificially induce tree dormancy by defoliating after harvest (Griesbach, 2007). Application of hydrogen cyanamide spray has been effective in breaking dormancy, thereby promoting bloom (Chabchoub *et al.* 2010). Traditional agronomical methods can be adopted in cold temperature mitigation, e.g., mulching, smoking and fogging, wind protection through crop framing or supporting by wooden support, fruit bagging, pruning and trimming, drip irrigation etc. Gupta *et al.* (2014) recorded maximum soil organic carbon (SOC) stock under lemon orchard (38.92 t/ha), followed by mango (33.11 t/ha), ber (29.28 t/ha), aonla + *aloe vera* (27.74 t/ha), guava (27.70 t/ha), kinnow mandarin (21.90 t/ha), phalsa (21.74 t/ha and the minimum SOC stock was under bael orchards (18.55 t/ha). Policy and institutional capability development should also be included in the adaptation and mitigation programmes. Encouragement to reduce greenhouse gas emissions should be made for horticultural, processing and marketing activities through tax preference, grant provision, free technical training or assistance (Diengngan and Hasan 2014). It is desirable to collect mulitilocational database on the basis of tree phenology, coupled with weather records since fruit crops are in urgent need for climate change adaptation. More research should be made in breeding, photosynthesis, biological nitrogen fixation, stress resistance, protected horticulture and precision horticulture for the enhancement of our adaptability to climate change (Dey *et al.* 2016).

13.12 Effect of Climate Change on Temperate Fruit Crops

Winter chilling is essential for most of the plants that fall dormant in the winter in order to avoid frost damage and do not resume growth until a certain amount of winter chill has accumulated for fulfilling their chilling requirement (Rai *et al.* 2015). Climate change is likely to affect chilling requirement of temperate fruit crops significantly and therefore, the opportunity to meet this requirement will be reduced as the climate becomes warmer. Increase in average global temperature would move the existing plant species and varieties to new latitudinal belts with favourable climates. It is, therefore, possible that crops that are used to be productive in one area may no longer be so or the other way round. The resultant of these climate changes is clearly apparent in the shifting of apple cultivation from lower elevations to higher altitudes in India. Successful cultivation of many fruit and nut trees requires the fulfilment of certain chilling hour for attainment of flowering and lack of chilling hour can delay or check flowering which may lead to spread out bloom, and cause various forms of abnormal growth. Climate change is likely to affect chilling (Luedeling *et al.* 2011). The global warming caused loss of vigour, fruit bearing ability, reduction in size of fruits, less juice content, low colour, reduced shelf life and increasing attack of pests resulting in the low production and poor-quality apple crop (Jangra and Sharma 2013). Vulnerability, rarity and rapid extinction of many species of temperate fruits will be among the other consequences of climate change.

In apple and pear orchards, some changes that occur due to climatic alteration (Dey *et al.* 2016) are mentioned below:

i) Possible increase in yield may be due to more carbon dioxide available for growth. But this may be limited by the availability of water and nutrients due to occurrence of higher temperature. Earlier and quicker ripening of fruits may occur with an increase in average summer temperature.

ii) Apple trees need several weeks of cold weather to complete dormancy. Cultivars requiring a longer dormancy may develop 'blind buds' that fail to develop in spring. Sporadic flowering over a prolonged period may lead to pollination problems. Prolonged wet soil conditions, poor or slow drainage will increase root losses and the risk of tree death from water-logging or crown rot caused by *Phytophthora* spp. Wet soil conditions increase lodging risk in windy weather.

iii) Apple trees can tolerate short spells of drought but prolonged dry weather can lead to root stress, defoliation, premature fruit drop or poor fruit size. Some cultivars of apple can suffer from sun scorch in very hot weather resulting in misshapen and unmarketable fruit. Warmer summers may see an increase in pest damage through some pests having more generations per year; minor pests becoming major pests, and through the introduction of foreign pests (Anon 2008). High temperature and moisture stress also increase sunburn and cracking in apples, apricot and cherries. The production of apple has gradually increased but the productivity has fallen from 10.8 to 5.8 t/ha (Awasthi *et al.* 2001). Among all the factors reducing productivity, climate is difficult to manage. The changes in climate in the form of erratic precipitation, increase in temperature, lesser days serving as the chilling period have started affecting the mountain agricultural production systems and ultimately the food security of the people (Datta 2013). According to Rai *et al.* (2015), insufficient chilling greatly influences flower initiation and fruit colouration along with deterioration in fruit texture and taste. Further, the lack of proper chilling is also posing serious problems like scab disease, premature leaf fall and infestation of red spider mite in apple. High temperature and moisture stress are increasing sunburn and cracking in apple, apricots and cherries in the higher altitudes. Insufficient chilling reduces pollination, fruit set and ultimately the yield in walnuts, pistachio and peaches. Advanced flowering has been found in olive, apple and pear. Reduced flower size and pedicel length were observed in cherry due to less chilling.

Cool temperatures in the winter are essential for fulfilling their chilling requirements to ensure homogeneous flowering and fruit set and generate economically sufficient yields. In order to escape the damage of sensitive tissues from winters, trees from temperate or cold climates have evolved the mechanism of dormancy. After certain duration of cold conditions (chilling), endo-dormancy is broken and the tree is ready to resume growth in the following spring (Luedeling 2011). Chilling requirement is defined as the number of effective chilling hours needed to restore bud growth potential in spring (Richardson *et al*. 1974). The chilling requirement is typically measured in terms of numbers of hours, during which temperature remains at or below 7°C during the winter season. The chilling requirements of different temperate fruits are presented in Table13.2.

Table 13.2: Chilling requirements (hours) of temperate fruits

Fruit crops	Chilling hours
Apple	1000- 1500
Pear	1200
Peach	200-1200
European plum	1000-1200
Japanese plum	700-1000
Cherry	2000-2700
Almonds	800
Apricot	300-900
Walnut	200-800
Pecan nut	400
Pistachio nut	700-1000

Source: Rai *et al*. (2015)

Dormancy is a mechanism that plants use to protect sensitive tissue from unfavourable climatic condition (Campoy *et al*. 2011). The rapid climatic changes might alter the adaptability of many temperate fruit crops in the near future, and severe productivity problems might arise. Commercially successful cultivation of many fruit and nut trees requires the fulfilment of a winter chilling requirement, which is specific for every tree cultivar. Lack of chilling as in mild winter conditions result in abnormal pattern of budbreak and development in temperate fruit trees (Ameglio *et al.* 2000). In temperate climate areas, frost can also represent the main cause of weather-related damage to crops. Apple and other temperate fruits are vulnerable to spring (late) frosts. During the bloom stage, a single event with temperatures going a few degrees below zero is sufficient to damage flower buds or even kill them. While light frosts result in the deterioration of fruit quality, severe frosts threaten the harvest itself (Snyder and de Melo-Abreu 2005). For fruits that are cross-pollinated such as walnuts and pistachios, insufficient chilling can reduce pollination leading to reduced crop yields (Gradziel *et al.* 2007). The optimum temperature for pollination and fertilization in temperate fruits like apple, pear, plum, cherry etc. is between 20-25°C. Low temperatures and rainy or foggy conditions had been observed to have a negative effect during pollination in sour cherry in USA (Zavalloni *et al.* 2008). High temperature in spring results in faster reproduction rate, thus the increase in pest population is rapid. According to Moretti *et al.* (2010), temperature variation can directly affect crop photosynthesis, and a rise in global temperature can be expected to have significant impact on the postharvest quality by altering important quality parameters such as synthesis of sugars, organic acids, antioxidant compounds, peel colour and firmness. The climate change affects not only the winter chilling of fruit crops but it also affects the other aspects like increase in the incidence of physiological disorders, pollination failure and phenology. (Rai *et al.* 2015).

13.12.1 Manipulation of The Chilling Requirement of Temperate Fruit Trees

1. Introduction and adaptation of low chilling cultivars of crops like apple, peach, pear and plum in certain areas of lower hills and North Indian plains where they could be grown commercially (Table13.3).

Table13.3 Low chill cultivars of some temperate fruits

Fruit crops	Low chill cultivars	Chilling requirement
Apple	Anna, Mayan, Tamma, Vered, Tropical Beauty, Parlin's Beauty, Schlomit, Michel, Neomi.	<800 hours
Peaches & Nectarines	Flordasun, Flordared, Sunred Nectarine, Sun Gold, Saharanpur Prabhat, Shan-e-Punjab, Sharbati	<500 hours
Pear	Gola, Pathernakh (Sand Pear), Leconte, Kieffer, Punjab Nectar.	150 hours
Plum	Santa Rosa, Jamuni, Alubokhara, Alucha purple, Titron, Satluj purple, Kala Amritsari.	---
Apricot	New Castle, Early Shipley, St.Ambroise, Kaisha, Chaubattia Alankar.	---

Source: Rai *et al.* (2015)

2. Evaporative cooling helps in reducing the bud temperature under mild winter condition and thereby increasing the number of chilling hours required for proper bud burst. Sprinkling with water to provide evaporative cooling during rest period had been found to advance the blooming of 'Flordagold' peach and 'Sungold' nectarine by 7 days (Gilreath and Buchanan 1979). Overhead irrigation has successfully been applied in Israel for cooling. Temperature is often considered the most important factor influencing phonological phases of fruit trees in temperate climates.
3. Chandler (1960) showed that bud burst was induced in apple trees when they were exposed to 44-46°C for six hours of a single or on two consecutive days in July, October and November.
4. Insufficient chilling during winter period usually results in delayed and erratic blooming and foliation of deciduous fruit and nut trees (Crane and Takeda 1979). The rest can be broken under certain conditions by using chemicals such as DNOC oil (Samish1945) or cyanamide (Shulman *et al.* 1983). Environmental conditions during the rest period are especially important in determining the optimum spray time of rest-breaking agents.

13.13 Impact of Climate Change on Plantation Crops

Exposure of coconut seedling to intense light intensity stress may cause seedling mortality. Dropping/wilting of leaves and death of coconut seedlings may occur if water deficit condition prevails during prolonged rain free periods. Decreased production of female flowers, reduced nut setting and higher button shedding may ultimately reduce nut size and yield in coconut. In Cashew high temperature (>34.4°C) and low relative humidity (<20%) during afternoon cause drying and dropping of flowers, resulting in poor fruit set and yield. Total yield loss up to 50 to 65% was reported in plantation crops.

13.14 Impact of Climate Change on Spice Crops

High temperature and decrease in the annual rainfall resulted in poor productivity trend in most of the black pepper growing areas of India. Pollination will be affected adversely because of higher temperature. Flower and fruit drop will occur frequently in cardamom and black pepper under high temperature condition. Saffron crops have suffered a 40% drop in production due to flower drop because of higher temperature (28°C). Higher temperature (33 to 35°C), in general, increases the rate of plant maturity in annual seed spices, thus shortening the growth stages.

13.15 Effect of Climate Change in Vegetable Crops

The importance of vegetables in providing balanced diet and nutritional security has been realised throughout the world. Vegetables are now recognized as protective food as they are good sources of vitamins and minerals and also play an important role in overcoming micronutrient deficiencies and provide opportunities of higher farm income. Vegetables are generally sensitive to sudden climate change condition. Higher temperature, reduced irrigation water availability, flooding, and salinity are the major limiting factors in sustaining and increasing vegetable productivity. Extreme climatic conditions will also negatively impact soil fertility and increase soil erosion. Tomatoes are strongly modified by temperature alone or in conjunction with other environmental factors (Abdalla and Verkerk 1968). High temperature stress disrupts the biochemical reactions fundamental for normal cell function in plants. It primarily affects the photosynthetic functions of higher plants. High temperatures can cause significant losses in tomato productivity due to reduced fruit set, and smaller and lower quality fruits. Symptoms causing fruit set failure at high temperatures in tomato include bud drop, abnormal flower development, poor pollen production, dehiscence, and viability, ovule abortion and poor viability, reduced carbohydrate availability, and other reproductive abnormalities and consequently lower productivity. Water availability is expected to be highly sensitive to climate change and severe water stress conditions will affect crop. Water availability is expected to be highly sensitive to climate change and severe water stress conditions will affect crop productivity, particularly that of vegetables. Vegetable production occurs in both dry and wet seasons in the tropics. However, production is often limited during the rainy season due to excessive moisture brought about by heavy rain. Most vegetables are highly sensitive to flooding and genetic variation with respect to this character is limited, particularly in tomato. The severity of flooding symptoms increases with rising temperatures; rapid wilting and death of tomato plants is usually observed following a short period of flooding at high temperatures.

Strategies to mitigate the adverse effect of climate change in vegetables include modifying fertilizer application to enhance nutrient availability to plants, direct delivery of water to roots (drip irrigation), grafting to increase flood and disease tolerance, and by use of soil amendments to improve soil fertility and enhance nutrient uptake by plants etc. Various crop management practices such as mulching and the use of shelters and raised beds help to conserve soil moisture, prevent soil degradation, and protect vegetables from heavy rains, high temperatures, and flooding. The use of organic and inorganic mulches is common in high-value vegetable production systems. These protective coverings help reduce evaporation, moderate soil temperature, reduce soil runoff and erosion, protect fruits from direct contact with soil and minimize weed growth.

References

Abdalla AA and Verderk K (1968). Growth, flowering and fruit set of tomato at high temperature. *The Neth. J. Agric. Sci.*, **16**: 71-76.

Aggarwal PK, Joshi HC, Singh SD, Bhatia A, Jain N, Shiv Prasad, Chaudhary A, Gupta N and Pathak H (2009). Agriculture and environment, In: *Hand Book of Agriculture*, Directorate of Information and Publication Agriculture, ICAR, New Delhi, pp. 62-92.

Alfonso DR and Brent MS (2014). Agricultural adaptation to climate change in the Sahel: expected impacts on pests and diseases afflicting selected crops, *African and Latin American Resilience to Climate Change Project*, USAID, Washington. pp: 53-54.

Ameglio T, Alves G, Bonhomme M and Cochard HE (2000). Winter functioning of walnut: involvement in branching processes. Pp.230-238. *In: L'Arbre, Biologieet development,* Montreal (CAN), Isabelle Quentin.

Anonymous (2001). Observed changes in climate and their effects. *In: Core Writing Team,* IPCC.

Anonymous (2008). *In: Herefordshire Orchards Community Evaluation Project*, UKCIP, 2008.

Awasthi RP, Verma HS, Sharma RD, Bhardwaj SP and Bhardwaj SV (2001). Causes of low productivity in apple orchards and suggested remedial measures. *In. Productivity of Temperate Fruits. Solan: Dr. Y.S. Parmar University of Horticulture and Forestry.* Jindal KK, Gautam DR, editors. 1–8.

Campoy JA, Ruiz D and Egea J (2011). Dormancy in temperate fruit trees in a global warming context: a review. *Sci Hortic.*, **130**: 357-372.

Chabchoub MA, Aounallah MK and Sahli A (2010). Effect of hydrogen cyanamide on bud break, flowering and fruit growth of two pear cultivars (*Pyrus communis*) under Tunisian condition. *Acta Horticulturae*, **884:** 427– 432.

Chandler WH (1960). Some studies of rest period in apple trees. *Proc. Amer. Soc. Hort. Sci.*, **76**: 1-10.

Crane J and Takeda F (1979). The unique response of the pistachio tree to inadequate winter chilling. *Hort Sci.*, **14:** 135-137.

Datta S (2013). Impact of climate changes in Indian Horticulture- A review. *Int. J. Science, Environment and Tech.*, **2** (4): 661– 671

Dey K, Ghosh K, Priya B, Dutta P and Das S (2016). Impact of climate change and mitigation strategies on fruit production. *Adv. Life Sci.*, **5** (7): 2588-2596.

Diengngan S and Hasan AM (2014). Impact of climate change on fruit crops. *Paper Presented in the International Conference on Horticulture for Nutritional, Livelihood and Environmental Security in Hills: Opportunity and Challenges* from May 22-24, 2014 held at Kalimpong, Darjeeling, West Bengal, India.

Gilreath PR and Buchanan DW (1979). Evaporative cooling with overhead sprinkling for rest termination of peach trees. *Proc. Fla. State Hort. Soc.*, **92**: 262-264.

Gradziel TM, Lampinen B, Connell JH and Viveros M (2007). 'Winters' almond: An early-blooming, productive, and high-quality pollinizer for 'Nonpareil'. *Hort. Sci.*, **42**: 1725-1727.

Griesbach J (2007). *In: Growing Temperate Fruit Trees in Kenya.* World Agroforestry Center (ICRAF), Nairobi, Kenya.

Gupta MK, Sharma SD and Kumar M (2014). Status of sequestered organic carbon in the soils under different land uses in southern region of Haryana, *International Journal of Science, Environment and Technology*, **3** (3): 811-826.

Hatfield JL and Prueger JH (2015). Temperature extremes: Effect on plant growth and development, *Weather and Climate Extremes*, **10** (4): 4-10.

Hazarika TK (2013). Climate change and Indian horticulture: opportunities, challenges and mitigation strategies. *Int. J. Environmental Engineering and Management*, **4** (6): 629-630.

Hemanth KN, Ravishankar KV, Narayanaswamy P and Shivashankara KS (2008). Influence of temperature on spongy tissue formation in Alphonso mango. *Int. J. Fruit Sci.*, **8** (3): 226-234.

Henson R (2008). In: *The Rough Guide to Climate Change* (2nd Ed.). Penguin Books, London, p. 384.

IPCC (2001). Climate change 2001: impacts, adaptation, and vulnerability. In: McCarthy J, Canziani O, Leary N, Dokken D and White K (ed.) *Intergovernmental Panel on Climate Change (IPCC),* Cambridge: Cambridge University Press.

IPCC (2007). *Climate Change: The Scientific Basis*. Oxford University Press, Oxford.

IPCC (2007a). Climate change 2007: The physical science basis. *In: Contribution of Working Group I to the Fourth Assessment Report of the Intergovernmental Panel on Climate Change.* Cambridge University Press, Cambridge, UK and New York, USA, pp. 996.

Issarakraisila M, Considine JA and Turner DW (1991). Pattern of vegetative and reproductive growth of mango trees in warm temperature region of Western Australia. *Acta Horticulturae*, **291**: 188-197.

Jangra MS and Sharma JP (2013). Climate resilient apple production in Kullu Valley of Himachal Pradesh. *Int. J. Farm Sci.*, **3**: 91-98

Katrodia JS and Sheth IK (1989). Spongy tissue development in mango fruit of cultivar alphonso in relation to temperature and its control, *Acta Hort.*, **231:** 827-834.

Kays SJ (1997). In: *Postharvest Physiology of Perishable Plant Products.* Athens, Avi. p.532.

Lloyd J and Farquhar GD (2008). Effects of rising temperatures and (CO) on the physiology of tropical forest trees. *Philosophical Transactions of the Royal Society of Biological Sciences*, **363**: 1811–1817.

Luedeling E, Girvetz EH, Semenov MA and Brown PH (2011). Climate change affects winter chill for temperate fruit and nut trees. *PLOS ONE*, **6**: e20155.

Mark DCJ and Marin RA (2016). Carbon sequestration potential of fruit tree plantation in southern Philippines, *J. Biodiversity and Environmental Sci.*, **8** (5): 164-174.

McCarty JJ, Canziani OF, Leary NA, Dokken DJ and White KS (2001). *Climate change 2001*. Impacts, adaptation and vulnerability. Contribution of Working Group II to the third Assessment Report of the IPCC. Cambridge University Press, Cambridge, UK.

Misselhorn A, Aggarwal P, Ericksen P, Gregory P, Horn-Phathanothai L, Ingram J and Wiebe K (2012). A vision for attaining food security', *Current Opinion in Environmental Sustainability,* **4** (1): 7-17.

Moretti CL, Mattose LM, Calbo AG and Sargent S (2010). Climate changes and potential impacts on postharvest quality of fruit and vegetable crops: A review. *Food Research International.*, **43:** 1824-1832.

Newman JA (2004). Climate change and cereal aphids: the relative effects of increasing CO_2 and temperature on aphid population dynamics, *Global Change Biol.*, **10**: 5-15.

Pathak H, Bhatia A, Jain N and Aggarwal PK (2010). Greenhouse gas emission and mitigation in Indian agriculture - A review, In: *ING bulletins on regional assessment of reactive nitrogen, SCON-ING,* New Delhi. **19:** 1-34.

Ploetz R and Ploetz R (2003). In: *Diseases of Tropical Fruit Crops.* CABI Publishing. pp. 327-363.

Rai R, Joshi S, Roy S, Singh O, Samir M and Chandra A (2015). Implications of changing climate on productivity of temperate fruit crops with special reference to Apple. *J. Hort.*, **2** (2):1-6. DOI: 10.4172/2376-0354.1000135

Rajatiya J, Varu DK, Gohil P, Solanki M, Halepotara F, Gohil M, Mishra P and Solanki R (2018). Climate change: impact, mitigation and adaptation in fruit crops. *Int. J. Pure App. Biosci.,* **6** (1): 1161-1169.

Richardson EA, Seeley SD and Walker DR (1974). A model for estimating the completion of rest for Redhaven and Elberta peach trees. *Hort. Sci.*, **9:** 331–332.

Samish RM (1945). The use of dinitro cresol mineral oil sprays for the control of prolonged rest in apple orchards. *J. Pomol. Hort. Sc*i., **21**: 164-179.

Shulman Y, Nir G, Fauberstein L and Lavee S (1983). The effect of cyanamide on the release from dormancy of grapevine buds. *Sci Hort*., **19**: 97-104.

Snyder RL and de Melo-Abreu JP (2005). Frost Protection: fundamentals, practice and economics, Vol. 1. *Environmental and Natural Resources Series*, FAO, Rome, p. 83

Staley JT and Johnson SN (2008). Climate change impacts on root herbivores. In: *Root Feeders: An Ecosystem Perspective* (eds. SN Johnson and PJ Murray). Wallingford, UK: CABI.

Trostle R (2008). Fluctuating food commodity prices - a complex issue with no easy answers. *Amber Waves,* **6** (5): 10-17.

Varu DK, Viradia RR, Chovatia RS and Barad AV (2010). Response of different genotypes of custard apple to weather parameters, *AGRESCO Report-2010*, JAU, Junagadh. pp. 24-36.

Varu DK and Viradia RR (2015). Damage of mango flowering and fruits in Gujarat during the year 2015, Survey Report of Department of Horticulture, JAU, Junagadh.

Venkateswarlu B, Maheswari M, Srinivasa Rao M, Rao VUM, Srinivasa Rao Ch, Reddy KS, Ramana DBV, Rama Rao CA, Vijay Kumar P, Dixit S and Sikka AK. (2013). *National Initiative on Climate Resilient Agriculture (NICRA), Research Highlights (2012-13).* Central Research Institute for Dryland Agriculture, Hyderabad.

Whiley AW and Schaffer B (1997). Stress physiology. In: *The Mango: Botany, Production and Uses*. Litz RE, editor. CAB International, 198 Madison Avenue, New York. pp. 147-174.

Zavalloni C, Andresen JA, Black JR, Winkler JA and Guentchev G (2008). A preliminary analysis of the impacts of past and projected future climate on sour cherry production in the Great Lakes Region of the USA. *Acta Hort*., **803:** 123-130.

Zekri M (2011). Factors affecting citrus production and quality, Citrus industry. Available at *crec.ifas.ufl.edu.*

Outcomes Assessment

Part A

Answer the following questions true or false.

1. CO_2 is the minor contributor among the greenhouse gases causing global warming. True/False
2. Changed climatic parameters disturb the flowering pattern in fruit crops. True/False
3. ‘Anna’ is a low chill cultivar of apple. True/False
4. Evaporative cooling does not help in reducing the bud temperature under mild winter condition. True/False
5. Due to high temperature several physiological disorders of fruit crops are more pronounced. True/False

Part B

Answer the following questions.

1. What is dormancy?
2. Cite two low chill cultivars of apple.
3. — and — are the greatest concern of mankind in 21st century.
4. Climate changes have negative/positive impact on tropical regions.
5. Insufficient chilling during winter period usually results in — and — blooming.

Part C

Write a brief note on each of the following.

1. Mitigation of climate change effect in fruit crops.
2. Effect of climate change on temperate fruit crops.
3. Manipulation of the chilling requirement of temperate fruit trees.
4. Effect of climate change in vegetable crops.
5. Adaptation and mitigation strategies.